GENERAL CHEMISTRY I WORKBOOK

ALPHEUS MAUTJANA

FOR YOUR OWN READING and PRACTICE

Kendall Hunt
publishing company

Cover image © Shutterstock, Inc.

Running Header Images (left) © Mr. Rashad/Shutterstock.com and (right) © AVIcon/Shutterstock.com

www.kendallhunt.com
Send all inquiries to:
4050 Westmark Drive
Dubuque, IA 52004-1840

Copyright © 2020 by Kendall Hunt Publishing Company

ISBN: 978-1-7924-2926-2

All rights reserved. No part of this publication may be reproduced, stored in a retrieval system, or transmitted, in any form or by any means, electronic, mechanical, photocopying, recording, or otherwise, without the prior written permission of the copyright owner.

Printed in the United States of America

CONTENTS

Chapter 1 Measurement **1**
Units of Measurements 1
Calibration and Significant Figures 2
The Steps of Making a Measurement 4
Unit Conversion and Calculations 6
"Moles" of a Chemical Substance 7
Lab 1: Taking Measurements 9

Chapter 2 Identities of Chemicals **21**
Chemical Species 21
Chemical Compounds 22
Chemical Symbols 26
The Naming of Chemicals 28
Lab 2: Application of Chemistry Skills 32

Chapter 3 Quantities of Chemicals **45**
Compound Stoichiometry 45
Solution Stoichiometry 50
Inventory of Solutions 52
Lab 3: Stoichiometry of Chemical Compounds 56

Chapter 4 Chemical Equations **67**
Molecular (or Reagent) Equations 67
Total and Net Ionic Equations 69
Reaction Stoichiometry 69
The Limiting Reactant 70
Balancing Chemical Equations 71
Precipitation Reactions 73
Lab 4: Stoichiometry of Chemical Reactions 75
Reduction–Oxidation Reactions 81
Oxidation Number 82
Acid–Base Reactions 84
Lab 5: Acid–Base Titration 87

Chapter 5   The Atom and Its Internal Structure   97

Development of the Atomic Theory   97

Atomic Orbitals and Energy Levels   98

Quantum Numbers   99

Electron Configuration   100

Periodic Trends   103

Lab 6: Reduction–Oxidation (Redox) Reactions   105

Chapter 6   Molecular Structures   117

Lewis Symbols and Chemical Bonds   117

Building Lewis Structures   119

Resonance, Formal Charge, and Isomers   122

Three-Dimensional (3D) Shapes and Drawings   124

Polarity of Molecules   130

Hybridization of Orbitals   131

Properties of a Covalent Bond   133

Chapter 7   Gases and Condensed Phases   143

The Behavior of Gases   143

Individual Gas Laws   143

The Combined Gas Law   145

The Ideal Gas Law   146

The Law of Partial Pressures   148

Lab 7: Discovering Physical Laws of Gases   150

Intermolecular Forces   161

Solution Formation   162

Colligative Properties   163

Lab 8: Forces that Hold Molecules   167

Chapter 8   What is Thermodynamics?   179

Flow of Energy   179

Thermochemical Equations   181

Calorimetry   183

Bond Energy   185

Hess' Law   187

Standard Enthalpies of Formation ($\Delta H_f°$)   188

Entropy Change (ΔS_{rxn})   189

Free Energy Change, (ΔG_{rxn})   191

Lab 9: Measuring Change in Energy   194

MEASUREMENT

Units of Measurements

Mathematics expresses ideas using *exact numbers* which are agreed upon or obtained by counting. Exact numbers have no *errors*. Over time, artifacts representing units of different types of quantities were developed. Seven base units have been established, known as the international system of units, *abbreviated from French* as "SI" units. These are *meter (m)* for distance, *Kelvin (K)* for temperature, *kilogram (kg)* for mass, *ampere (A)* for electric current, *candela (cd)* for luminosity, *moles (mol)* for chemical species, and *second (s)* for time. From these we can derive units of other quantities such as cm^3 for volume. Size of a given quantity can be represented by a number including fractions of the applicable unit. That *number* along with a symbol of the *unit* is called a *measurement*.

Since specific quantities such as 1 in, 1 kg, 1 mile, etc. as well as their equivalents, that is, 2.54 cm, 2.20462 lb, 1.60934 km, etc. have been agreed upon (or defined), they can be regarded as _____ numbers—that is, they have no error associated. In contrast, measured quantities have _____ associated.

The seven base quantities and units are:

Quantity type	Units
1. luminosity	cd
2. _____	_____
3. _____	_____
4. _____	_____
5. _____	_____
6. _____	_____
7. _____	_____

© fosgen/Shutterstock.com

My Professor Says:

Do I get it? Let me check:

For very small or very large measurements, we use *prefixes to represent a fraction or multiple* of the unit, expressed in scientific notation, that is, exponents of 10. For example, *1 mol/1000 is 1×10^{-3} mol or 1 mmol* while *1000 mol is 1×10^{3} mol or 1 kmol*.

I know a prefix can be used with any unit, or be replaced with $\times 10^-$ from the table. Let me fill in the missing prefix or exponent:

a) 2 ___ A = $1 \times 10^{\underline{9}}$ A (or 1 000 000 000 A)

b) 8.4 ___ s = $8.4 \times 10^{\underline{-3}}$ s (or 8.4/1000 s)

c) 1 kg = 1×10—— g (or 1000 g)

d) 1 μmol = 1×10—— mol (or 1/1 000 000 mol)

PREFIX	tera	giga	mega	kilo	m (meter)	deci	centi	milli	micro	nano	pico
SYMBOL	T	G	M	k	m	d	c	m	μ	n	p
NUMBER	10^{12}	10^{9}	10^{6}	10^{3}	10^{0}	10^{-1}	10^{-2}	10^{-3}	10^{-6}	10^{-9}	10^{-12}

Calibration and Significant Figures

Analog measuring devices have calibration marks which are numbered according to increasing magnitudes, called a scale. For example, tick-marks on a meter-stick represent a scale—*analog* scale—which always has spaces between the markings, regardless of how close they are. To make a measurement, we put the item being measured against the scale (or *vice versa*) and read the markings. Electronic (also called digital) measuring devices display the number of units to indicate magnitude.

My Professor Says:

The marking next to the object's edge or to its actual magnitude represents "*certain digits*." We then add one last digit called the "*estimated digit*." This last digit represents the number of "imaginary" tenths in the space between scale markings. One user may imagine one-tenth more, or less, than another user. For this reason, every analog measurement has a margin of error or a level of *uncertainty* associated with it. This error margin or uncertainty *is* simply *give-or-take* ($\pm$) *a tenth of the smallest increment of the scale*. In contrast, the error margin on an electronic device is simply "*ones*" of the last decimal, give-or-take. The *certain* and *estimated* digits collectively make *significant figures*. A measurement always has the same number of decimals as *uncertainty* of the device. All the digits after leading zeros are *significant*. For example, *0.44 cm has 2 sig. figs., 0.02cm has 1, and 0.020 cm has 2*.

Speedometer;

_____

Do I get it? Let me check:

Practice Questions

1.1. Write the "uncertainty" for each device:

To do this, I divide the smallest increment by 10 (mark to mark, even if unnumbered):

Cylinder—measures liquid volume contained:

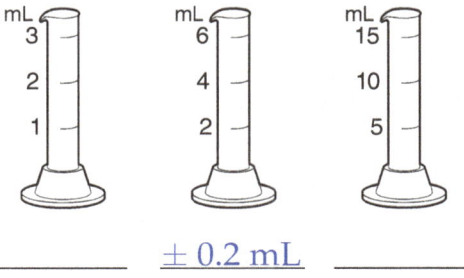

_____ $\pm$ 0.2 mL   _____

A buret—measures volume of liquid transferred out of it. So, it has zero at the top

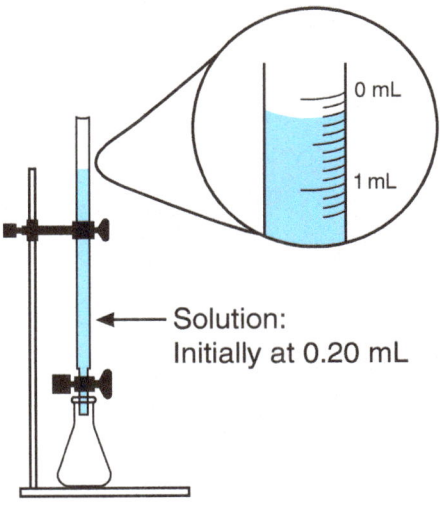

Solution: Initially at 0.20 mL

_____

3

My Professor Says:

© Mr. Rashad/Shutterstock.com

The Steps of Making a Measurement

We can illustrate the analog measurement using the following picture:

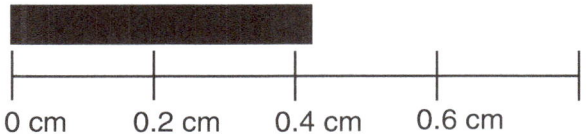

Clearly, the length of the object should be reported to two *significant figures—a certain digit* (*0.4 cm*) *and a second digit which we must estimate.* To make this measurement we add, to the length that is "*certain*" (0.4 cm), the *estimated* length of the part extending beyond it. We can follow these steps:

1. Note the certain digits (*0.4 cm* here)
2. Find the value of the smallest increment (*0.2 cm* here)
3. Divide smallest increment into 10 parts, that is, tenths (0.2 cm/10 = *0.02 cm*)
4. Count how many tenths cover the extending length (*2* tenths here).
5. Estimate the extending length as *2(0.02 cm)*; give or take a tenth of the smallest increment.
6. Add the estimated length to the certain length: *0.4 cm + 2(0.02 cm)*.

So, the object length is *0.44 cm ± 0.02 cm*.

Do I get it? Let me check:

© AVIcon/Shutterstock.com

Practice Questions

1.2. Make the following measurements:

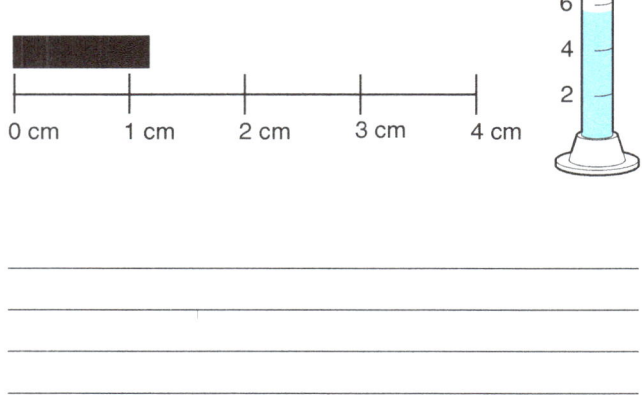

_____
_____
_____
_____
_____
_____
_____
_____

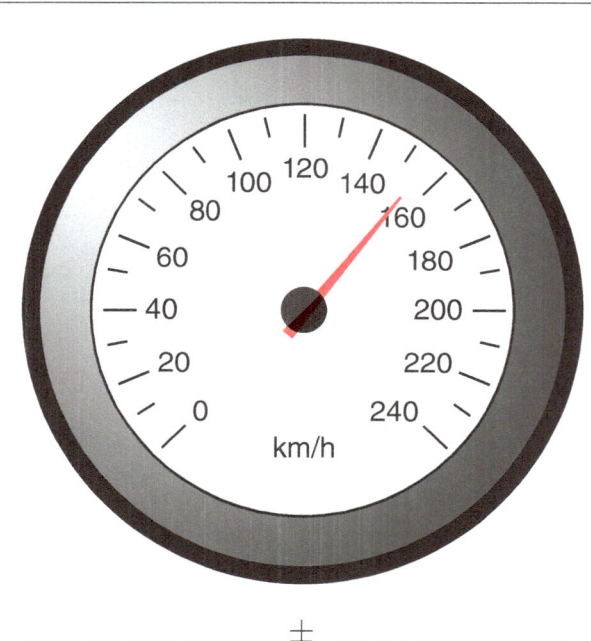

_____ ± _____

4

My Professor Says:

Do I get it? Let me check:

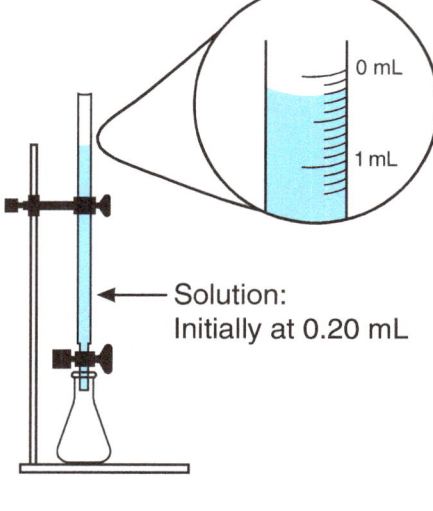

_____ ± _____   _____ ± _____

_____
_____
_____

Note that math operations (×, ÷, +, −, *power*) also apply to the units. We can follow the following rules:

1. In *multiplication* and *division*, the answer must have the same number of significant figures as the measurement with fewest significant digits, for example,

   $(234.506 \, cm) \times (4455.9 \, cm) \times (0.12 \, cm) = 1.3 \times 10^5 \, cm^3$

2. In *addition*, and *subtraction*, the answer must have the same number of decimals as the measurement with fewest decimals, for example,

   $4.506 \, cm + 0.12 \, cm + 455.9 \, cm = 460.5 \, cm$

Practice Questions

1.3. Give the final answer with units for each of the following measurement calculations.

It looks like I must find and use the applicable rule for each case.

a) 250. cm × **5.0** cm = _____

b) 0.**3** nm + 0.05 nm = _____

My Professor Says:

3. When rounding (let's say *to 3 sig. figs*):

   a) if the number removed is less than 5, leave the last digit unchanged, for example, *6073 becomes 6.07×10^3*.

   b) If the number being removed is greater than 5 or it is 5 followed by nonzero digits, increase the last significant digit by one unit, for example, *60,751 becomes 6.08×10^4*.

   c) If the number being removed is 5 followed only by zeros or nothing, *increase the last digit by one if it is odd* but leave it *if it is even*.

Unit Conversion and Calculations

Mathematical relationships between measurements allow conversion of one set of units to another using <u>dimensional analysis</u>. In this method, we generally multiply the measurement being converted by the "unit-equivalence." The unit-equivalence is whatever quantity that we know in the <u>desired units</u> over the same quantity in <u>units we are converting</u>. For example, *12.0 in can be converted to centimeters by multiplying it by 2.54 cm (desired units) over 1 in (units being converted). When we do, we get 30.5 cm (retain 3 sig. figs.).* The "unit-equivalence" is called a <u>conversion factor</u>. If necessary, we may have to multiply further by another *conversion factor* until we end with desired final units even if the desired units are compounded, for example, g/mL, mol/L, km/h.

Do I get it? Let me check:

c) $(1.5 \times 10^2 \, mg) \div (8.56 \, cm)^3 = $ _____

d) $\dfrac{1.529 \, atm \times (4.38 \times 10^2) \, L}{27.7 \, mol \times 294.15 \, K} = $ _____

Practice Questions

1.4. The carrot shown below is immersed in water and has a mass of 15.265 g. Determine its density in kg/m^3.

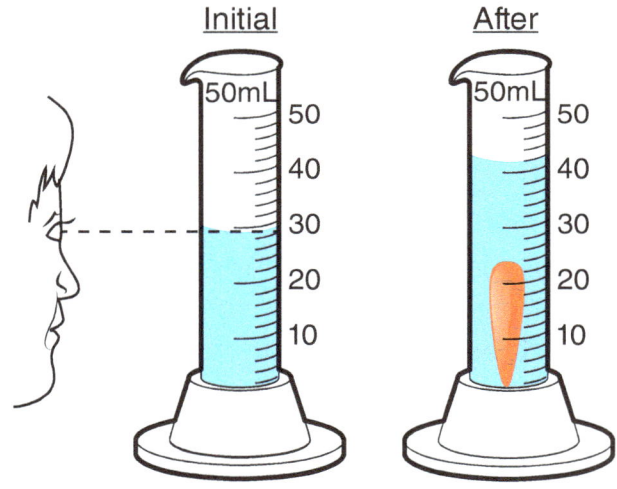

First of all, the uncertainty of the measuring cylinder is _____. Now, the number of decimals that my measurement should have is _____. I can read the initial volume of water which is _____. The final volume which includes the carrot is _____.

My Professor Says:

When we perform calculations, we always retain the uncertainty of the measuring device used. This is achieved by reporting the final answer with the same uncertainty.

"Moles" of a Chemical Substance

A mole is a fixed number of particles, the same way as a dozen, week, and so on. Exactly twelve eggs, or something else, make one dozen.

Exactly 7 days make one week. In the same way, 602 200 000 000 000 000 000 000 atoms or molecules make one mole. So, "mole" is a unit of measure for the amount of matter. To avoid mistakenly omitting or adding extra zero, the number of particles in one mole of substance is written in scientific notation as 6.022×10^{23} particles. This number is very large given the invisibly tiny sizes of chemical species, and it is named Avogadro's number in honor of Amedeo Avogadro, a scientist who studied the relationship between number of particles of a gas and the volume they occupy. We enter this number (and others in this notation) into a scientific calculator by typing:

Avogadro's number is simply *a number* and not a *math operation*. So, the *multiplication (×)* or *power (^)* signs are not needed when we enter it into a calculator.

Do I get it? Let me check:

I can now calculate the volume of the carrot as final minus initial volume which is
_____ − _____ = _____
_____.

To find density of the carrot, I must divide _____ by _____.
This gives me a density of _____.

I can now use the "units equivalence" (1 kg = 1000 g; 1 mL = 1 cm³, and 100 cm = 1 m with both sides cubed) as conversion factors until I end with kg/m³. Let me show my dimensional analysis set-up here:

I see that even if it is written in scientific notation, a number is still only a number, and not a math operation. So, Avogadro's number is simply a _____.

When using Avogadro's number in calculations, we should not enter it as if it has a _____ (×) or _____ (^) sign. The calculator function "**EXP**" or "**EE**" after the coefficient (6.022, in this case) means the base is ten (× **10—**) or 6.022 tens counted as many times as the exponent (23 times in this case).

THESE ARE MY OWN NOTES
(from listening to lectures, watching YouTube videos, etc.)

Date: _____

Date: _____

Date: _____

Date: _____

I wish the professor could explain this: _____

Student's Practical Laboratory Experiment
TAKING MEASUREMENTS

OBJECTIVE

The objective of this experiment is to demonstrate the proper use of different measuring devices for length, mass, and volume. Certainty of such measurements depends on the calibration of the device used and has direct effect on the *result* of an experiment. This will be demonstrated in the determination of density.

Images modified from © Shutterstock/chuhastock, udaix and Rvector

Measuring devices (left to right): Ruler, electronic balance, measuring cylinders, and a buret.

LEARNING OUTCOMES

After this experiment, students will be able to:

1. Determine the error margin (uncertainty) of a measuring device.
2. Include the estimated digit to the numbers representing the last calibration mark.
3. Apply the rules for using measured values in calculations.
4. Convert units of measurement.

SAFETY NOTES

Handle glassware with care and inspect it for cracks and sharp edges before use. Immediately wash off the chemicals should you make skin-contact. Always wear safety goggles and shoes when doing experiments. <u>When finished, wash your hands with soap and water.</u>

PRE-LAB ASSIGNMENT

Complete the following table for each of the measuring devices shown under objective (p. 9):

	Ruler	Units	Mass-Balance	35 mL Cylinder	70 mL Cylinder	Buret
Smallest increment						
Tenth of an increment						
Error margin (uncertainty)						
No. of decimals to report						

Materials Needed:

1. A 10 cm ruler
2. A calculator
3. A less than 1 cm width, cubic block/rectangular prism—one of different pure metal blocks
4. Aluminum (or other non–water-soluble) pellets
5. About 4 × 4 cm piece of aluminum foil
6. A 10 mL and a 100 mL graduated cylinders
7. An electronic mass-balance

EXPERIMENTAL PROCEDURE

Part I

Objects with a Regular Geometric Shape

Describe your cubic/rectangular block and use a ruler to measure its dimensions.

Physical Description	Dimensions (include units)	No. of Significant Figures
Appearance: _____	Length:	
_____	Width:	
Mass: _____	Height:	

Perform the calculations below. **Show your work**.

a) Use the formula for a rectangular prism to calculate the volume of your block.

b) Convert the volume of your block into ft^3. Report the answer in scientific notation. [2.54 cm = 1 in; 12 in = 1 ft]

c) Calculate the density $\left(\text{in } g/cm^3\right)$ of your block.

d) Convert the density of your block into kg/m^3.

Part II

Volume of a Liquid

Different glassware such as graduated cylinders, burets, pipets, as well as beakers, are designed to measure volume of a liquid. These have varying degrees of uncertainty depending on how coarsely or finely they are calibrated.

Your lab instructor has filled the 10 mL and 100 mL graduated cylinders as well as the 100 mL beaker with some distilled water. After setting the meniscus of the liquid to the zero ("0 mL") mark on the buret, the lab instructor transferred some liquid to a different container. Notice that the former set of glassware are calibrated "To Contain, TC," that is, they measure the volume of liquid contained inside. A buret and a pipet on the other hand, are calibrated "To Deliver, TD," that is, they measure the volume of liquid transferred or taken out.

Take a reading of the volume of water in each of these devices. Remember to read these volumes at the bottom of the meniscus, and to include estimated digit and units. It is useful to hold a piece of white paper behind the scale to make it clearer.

Glassware	Reading	Uncertainty
Beaker		
10 mL cylinder		
100 mL cylinder		
Buret		

Which of these devices provides the most *precise* measurement, that is, one closest to the true amount? (Hint: the smaller the uncertainty the higher the precision of the device)

Part III

Finding Density

To determine the density of a pure substance at a given temperature, we need its mass and volume. Powders and liquids must be weighed in containers. We always weigh the empty container first and again after adding the substance. We then subtract out the weight of the container. This is a technique called *"mass by difference."* If weighing a powder, you will need a weighing paper/boat.

1. Obtain a clean, small beaker. Weigh it and record its mass in the table below.
2. Transfer 1.1 to 1.5 g of aluminum pellets into the beaker.
3. Weigh and record the mass of the beaker and pellets in the same table. Calculate the mass of the pellets.

We have already seen that if we have mass of an object with a regular geometric shape, we can use a formula to find its volume. We can then calculate the object's density that way. If the object is irregularly shaped or subdivided, it is difficult to measure its dimensions in order to get its volume. We use a technique called *"volume by water displacement."* To get the volume of the pellets, we measure the volume of a liquid *before* and *after* the solid has been placed in the liquid. The difference in volume is the volume of the solid.

4. Pour 4–5 mL of water into your 10 mL graduated cylinder. Measure this volume precisely.
5. Carefully add all the previously weighed aluminum pellets to the water, making sure not to lose any water due to splashing. Also make sure that the pellets are all completely immersed in the water and that there are **no air bubbles**. Add pellets slowly to limit air bubbles.
6. Measure and record the new volume of the water plus the pellets.
7. When finished, dry the aluminum pellets and return them to the lab instructor.

	Mass (g)
Mass of empty beaker	
Mass of beaker and Al pellets	
Calculated mass of Al pellets	
Initial volume of water in cylinder	
Final volume of water and Al pellets	
Calculated volume of Al pellets	

14   General Chemistry I Workbook

To determine the density of aluminum pellets, we need the dry mass (measured above), and volume (measured by displacement).

8. Use your measured mass and volume of the aluminum pellets to calculate the density of aluminum.

Density of Al pellets, in g/cm^3. *Show your work below* and report your answer to the correct number of significant figures [1 mL = 1 cm^3]:

Instructor's **OK**: _____

–oOo–   **END OF LAB REPORT**

Measurement

THESE ARE MY OWN NOTES
(from listening to lectures, watching YouTube videos, etc.)

Date: _____

Date: _____

Date: _____

Date: _____

I wish the professor could explain this: _____

Measurement

IDENTITIES OF CHEMICALS

Chemical Species

In our normal day-to-day life, we experience *matter* (anything with mass and occupying space). Matter is found in three physical states—*gas*, *liquid*, or *solid*. This does not mean that every gas such as air, every liquid such as tap water, or every solid such as bread is a pure substance. Different gases can mix and make a gas mixture. The same is true of different liquids and different solid particles. A gas, liquid, or solid can be a *mixture*. Furthermore, a solid can mix with liquid or gas. At this level, things simply mix and become mixtures which can be classified as *homogeneous* (i.e., in one phase) or *heterogeneous* (i.e., in different phases). We can exploit physical properties of things that were mixed such as particle size, color, shape, and boiling temperatures and separate them by filtration, sorting, or distillation among other methods. If we limit our view to this level of gas, liquid, or solid, this is about as far as we will go. In this general chemistry course, we go deeper than that.

As we mentioned at the end of the previous chapter, a gas is made up of particles, so is a liquid and any solid. If all particles are of the same species, then that gas, liquid, or solid is a *pure substance*—even if some of those identical particles are clumped together in solid chunks while others are liquid. The *particles* that make matter can be *atoms* and *molecules*, or they can be *cations and anions*, *formula units*. We refer to these different types of particles as *chemical species*. Each of these will be discussed further.

In nature, there are different kinds of atoms called *elements* shown in the *Periodic Table* below–metals (shaded light green and orange) as well as nonmetals (dark blue). These elements are named, symbolized, and arranged systematically. *Atoms of the same element with the same number of protons but different number of neutrons* called *isotopes* ($^{A}_{Z}X$), also occur in nature. Atomic No. $Z = $ *no. of protons*; Mass No. $A = $ *protons + neutrons*.

My Professor Says:

© Mr. Rashad/Shutterstock.com

Do I get it? Let me check:

© AVIcon/Shutterstock.com

© MeKaDesign/Shutterstock.com

Chemical Compounds

The far-right column in the Periodic table consists of *Noble Gas* elements. These are the most stable, practically inert, because their outer shell of electrons is full, with two electrons ($2e^-$) for He or eight ($8e^-$) for other noble gas elements. It is very important to note that other elements want to have a full outer shell and be stable like noble gases. We say they want to have the *Noble Gas Configuration (NGC)*.

This picture shows that at room temperature, the physical state of salt is _____ and that of water is _____.
Salt dissolves in water and makes a _____ mixture. Pure oil does not dissolve in water. So these two make a _____ mixture.

I can label magnesium metal (**Mg**) and nonmetal, Sulfur (**S**) below. Each is a _____ substance with the same _____.

© Fablok/Shutterstock.com

_____

© pedphoto36pm/Shutterstock.com

_____

In the Periodic table, columns are called "Groups" and they are numbered 1A, 2A, (3B–8B, 1B, 2B) 3A, 4A, 5A, 6A, 7A, and 8A from left to right. Elements in the "A-series" are called main-Group elements while those in the "B-series" are

I can try to fill in the missing information:

Isotope	Atomic Number	Mass Number	Protons	Neutrons	Electrons
$_{20}$Ca		42			
	18			22	
			14	16	

22

My Professor Says:

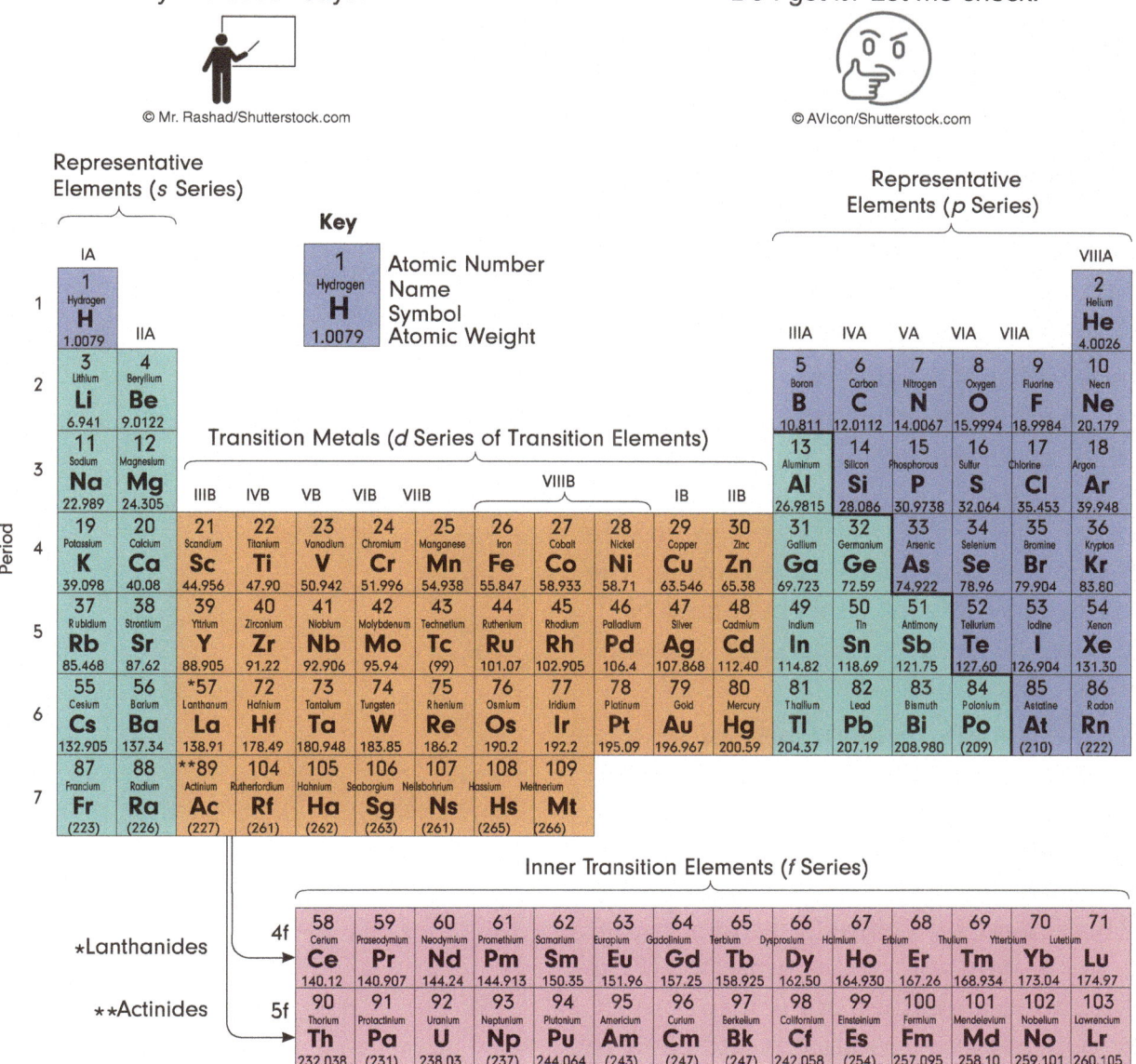

called transition metals. The Group number represents the number of electrons in the outer shell (also called *valence electrons*) of each element in that Group. The rows, numbered 1 (top) through 7 (bottom), are called "Periods" and the Period number represents the number of shells (also called *energy levels*) of each element in that Period. The element number (called *atomic number*) represents the number of positively charged protons in the nucleus of each element which keep negatively charged electrons

Do I get it? Let me check:

The _____ of **Mg** is **2A**, which means the number of electrons in its outer shell is _____. These are also called _____ electrons. To have _____ (NGC) of **Ne**, **Mg** will _____ its two valence electrons. Mg will not be neutral anymore, but it will have a _____ charge of _____.

My Professor Says:

Do I get it? Let me check:

electrostatically tethered, similar to how a moon is gravitationally tethered to the planet. Each element has the same number of electrons as its protons and therefore atoms are electrically neutral.

To achieve NGC, metal elements give away the few valence electrons they have (in their outermost shell) and remain with full shells. By doing so they become positively charged cations—one positive charge for each electron lost. For example, the element Na loses 1e⁻ and becomes Na⁺, Mg loses 2e⁻ and becomes Mg^{2+} and Al loses 3e⁻ becoming Al^{3+}.

Nonmetals on the other hand can achieve the NGC in one of two ways. One way is by receiving electrons from a metal element, and by doing so become negatively charged anions. The element Cl gains 1e⁻ and becomes Cl^-, S gains 2e⁻ and becomes S^{2-}, and P gains 3e⁻ to become P^{3-}. If the stable cation and anion come into contact, they become electrostatically attached to each other. By doing so metal cation and nonmetal anion form a formula unit, that is, "an ionic compound."

Another way by which nonmetals achieve NGC is sharing their unpaired electrons and getting covalently bonded to each other so that eight electrons (or 2 for hydrogen) can be counted around the nucleus of each. By doing so, nonmetals form a molecular compound, that is, "a molecule." Chemical species have unique identities.

On the other hand, the Group number of element **S** is _____. The element **S** needs _____ to have Noble Gas Configuration (NGC) of **Ar**. So, S will _____ two valence electrons. **S** will not be neutral anymore, but it will have a _____ charge of _____.

After **Mg** has lost its two valence electrons, it will be a _____ and the symbol we will use for it is _____

After **S** has gained two valence electrons, it will be an _____ and the symbol we will use for it is _____

An _____ compound is formed by electrostatic attraction of _____ and _____ ions.

My Professor Says:

Do I get it? Let me check:

Monoatomic Ions

IA (1)	IIA (2)	IIIB (3)	IVB (4)	VB (5)	VIB (6)	VIIB (7)	VIIIB (8)	VIIIB (9)	VIIIB (10)	IB (11)	IIB (12)	IIIA (13)	IVA (14)	VA (15)	VIA (16)	VIIA (17)
H^+																H^-
Li^+	Be^{2+}													N^{3-}	O^{2-}	F^-
Na^+	Mg^{2+}											Al^{3+}		P^{3-}	S^{2-}	Cl^-
K^+	Ca^{2+}				Cr^{2+} Cr^{3+}	Mn^{2+} Mn^{3+}	Fe^{2+} Fe^{3+}	Co^{2+} Co^{3+}		Cu^+ Cu^{2+}	Zn^{2+}				Se^{2-}	Br^-
Rb^+	Sr^{2+}									Ag^+			Sn^{2+} Sn^{4+}		Te^{2-}	I^-
Cs^+	Ba^{2+}												Pb^{2+} Pb^{4+}			

In addition to these simple ions consisting of a single element, there are polyatomic ions which consist of two or more elements. Formulas and names of common polyatomic ions are shown below:

Polyatomic Ions

HCO_3^-	hydrogen carbonate	HPO_4^{2-}	hydrogen phosphate	BrO^-	hypobromite	OH^-	hydroxide
CO_3^{2-}	carbonate	$H_2PO_4^-$	dihydrogen phosphate	BrO_2^-	bromite	CN^-	cyanide
NO_3^-	nitrate	PO_4^{3-}	phosphate	BrO_3^-	bromate	$C_2H_3O_2^-$	acetate
NO_2^-	nitrite	ClO^-	hypochlorite	BrO_4^-	perbromate	MnO_4^-	permanganate
HSO_4^-	hydrogen sulfate	ClO_2^-	chlorite	IO^-	hypoiodite	CrO_4^{2-}	chromate
SO_4^{2-}	sulfate	ClO_3^-	chlorate	IO_2^-	iodite	O_2^{2-}	peroxide
SO_3^{2-}	sulfite	ClO_4^-	perchlorate	IO_3^-	iodate	NH_4^+	ammonium
				IO_4^-	periodate		

My Professor Says:

Chemical Symbols

You might find it funny that the first 20 *elements* appear sequentially like a person's name and address:

H He**Li**Be**B** CNOF
Ne**Na** MgAlSiPS Cl
Ar**K**Ca

Notice that chemical symbols of the elements are either a single uppercase letter such as "**H**" for hydrogen, or an uppercase followed by a lowercase letter such as "**Li**" for lithium. However, the chemical symbols are not always abbreviations of English names but can be abbreviations of element names in other languages such as Latin or German. For example, the symbols "**Na**" and "**K**" come from German names "natrium" and "kalium" for sodium and potassium, respectively. While all metal elements in their pure form are represented by the symbols as they appear in the Periodic table, many nonmetal elements are found as molecules in their pure form as shown in the table below. Anions need cations to be stable:

Do I get it? Let me check:

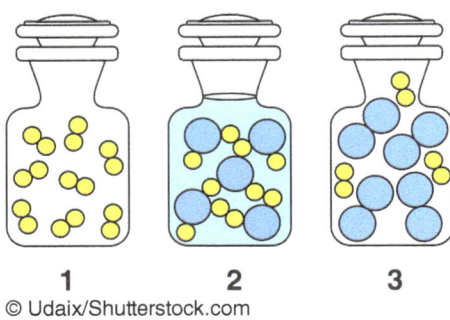

If the yellow particles are **H-atoms** and blue ones are **O-atoms**, then the particles in all these bottles are called _____ _____.

Bottle number _____ contains pure element, bottle number _____ contains pure _____, while bottle number _____ contains a mixture of _____.

We can write the chemical formula of molecules in bottle number 2 as _____

My Professor Says:

Element	Unstable Atom (not in this form in nature)	Stable ion (needs counter-ion)	Stable molecular Form
Hydrogen	H	# footnote	H_2
Nitrogen	N	N^{3-}	N_2
Oxygen	O	O^{2-}	O_2
Fluorine	F	F^-	F_2
Chlorine	Cl	Cl^-	Cl_2
Bromine	Br	Br^-	Br_2
Iodine	I	I^-	I_2

\# Hydrogen forms H^- ion but usually H^+ ion which is a proton

Using valence electrons, chemical elements can form two major types of compounds with each other—*ionic compounds* (i.e., *salts* by electrostatic attraction between cation and anion), as well as *molecular compounds* (i.e., molecules by sharing unpaired electrons between two or more atoms).

Do I get it? Let me check:

The formula of a compound formed by Mg^{2+} and S^{2-} ions is _____, and its chemical name is _____.

Ionic compounds are also called _____. The formula units stick to each other making a crystal lattice. Therefore, as shown below, salts occur as _____ in nature.

If the cation was Cr^{3+}, the formula of the compound would be _____, and the chemical name would be _____.

Magnesium Sulfide Model and Crystals

27

My Professor Says:

For an ionic compound, the sum of the positive charges must equal the sum of the negative charges. As illustrated above, one way of making sure there are enough cations to balance the anions or *vice versa*, is to make the numeric charge of the anion (usually a superscript) a subscript of the cation and, similarly, make the numeric charge of the cation a subscript of the anion. This gives the ionic compound symbol (called a *formula*).

The Naming of Chemicals

For this course you should memorize the symbols and names of the first 54 elements shown in the Periodic table on page 23. For simple ions of the main Group elements, the cation name is the same as the name of the neutral metal element but includes the word "ion." For example, Na^+ is sodium ion and Ca^{2+} is calcium ion while Al^{3+} is aluminum ion. Transition metal elements may form stable cations of different charges. For these we include a Roman numeral in parentheses after the metal name to identify the charge formed. For example, Cr^{2+} is chromium(II) ion while Cr^{3+} is chromium(III) ion. Simple anion names contain the root of the element name with the suffix "-*ide*" followed by the word "*ion*." For example, N^{3-} is nitr*ide* ion and O^{2-} is ox*ide* ion while F^- is fluor*ide* ion. See table (p. 25) for polyatomic ion names.

All ionic compounds consist of a cation and an anion which neutralize each other resulting in an electrically neutral

Do I get it? Let me check:

Practice Questions

2.1. Complete the following table:

I need to remember names, formulas, and charges of polyatomic ions and then switch them back to find ions and switch them over to make a compound.

Compound Formula	Cation Formula	Anion Formula	Name of Compound
Ca(OH)$_2$			
			Potassium nitrate
	Cr^{2+}	ClO$_3^-$	
Na$_2$O			
	Cu$^+$	CO$_3^{2-}$	
			Aluminum sulfide

My Professor Says:

formula unit. Compounds have no charge and should not be confused with polyatomic ions which do have overall charge. To name an ionic compound we simply separate the cation from the anion so that we can see its charge. When we separate the cation from the anion, we *take the subscript of the anion back to being the superscript of the cation* and, similarly, *take the subscript of the cation back to being the superscript of the anion*. Remember that equal subscripts are simplified into 1 each, and that polyatomic ions have their own subscripts which should not be made superscripts. So, it is important to remember the charge of every simple and polyatomic ion as shown on page 25. After splitting the ionic compound, we simply *name the cation followed by the anion*. For example, $MgBr_2$ separates into Mg^{2+} and Br^- and it is named magnesium bromide. The compound $AlPO_4$ separates into Al^{3+} and PO_4^{3-} and it is named aluminum phosphate, while Cu_2CO_3 separates into Cu^+ and CO_3^{2-} and it is named copper(I) carbonate.

In this course, we will classify molecular compounds into two categories—*binary covalent compounds* and *acids* which separate into H^+ as cation and the anion. For binary compounds, the *first element's name is used unchanged* with a Greek prefix other than mono-, and *the second element name is used* also with Greek prefix including "*mono-*" but *containing the root* of its original name with a suffix "*-ide.*" For example, N_2O_4 is named *di*nitrogen *tetr*ox*ide* while NO is nitrogen *mono*x*ide*. The Greek

Do I get it? Let me check:

Practice Questions

2.2. Write down the systematic names for the following molecular compounds:

I can do this just from remembering Greek prefixes—No mono for the first element.

a) SO_3 _____

b) P_4O_6 _____

c) SeO_2 _____

d) B_2S_3 _____

e) Cl_2O_7 _____

f) NS _____

29

My Professor Says:

prefixes are mono-(1), di-(2), tri-(3), tetra-(4), penta-(5), hexa-(6), hepta-(7), octa-(8), nona-(9), deca-(10), . . .

In this course, the chemical formula of an acid will always contain hydrogen (H) as the first element, occupying the same position as the cation in an ionic compound. All other hydrogens in the formula which are not written first, remain with the anion and do not separate as H^+ ion. To name a *binary acid*, we use the root "*hydro-*," and the root of the second element with the suffix "*-ic*" followed by the word "*acid*." For example, the acid HF(aq) is *hydrofluoric acid*. The notation "(aq)" for "aqueous" indicates that water is the solvent. If HF(g) is used, the notation "(g)" stands for "gas" in which case the appropriate name would be hydrogen fluoride since H behaves as if it is a cation "H^+" attached to the anion F^-). Another example of a binary acid is H_2S(aq) which is named *hydrosulfuric acid*. One exception, in the sense that it is not binary, is HCN(aq), which is also named *hydrocyanic* acid.

For oxyacids in which the anion is a polyatomic anion containing oxygen, we *use the root of the anion name* and replace suffixes "*-ate*" with "*-ic*" and "*-ite*" with "*-ous*" and add the word "*acid*." We do not use the root of hydrogen "hydro-" at all. For example, HOCN(aq) is named cyan*ic* acid (after cyanate ion, CNO^-), $HClO_4$(aq) is named *perchloric acid*, and HClO(aq) is named hypochlor*ous acid*.

Do I get it? Let me check:

Practice Questions

2.3. Complete the following table:

I need to remember names, formulas, and charges of polyatomic ions and then switch them back to find ions and switch them over to make a compound.

Acid Formula	Name of the acid
	Hydrosulfuric acid
HNO_2	
H_2SO_3	
	Carbonic acid
$HBrO_3$	
	Hypoiodous acid

THESE ARE MY OWN NOTES
(from listening to lectures, watching YouTube videos, etc.)

Date: _____

Date: _____

Date: _____

Date: _____

I wish the professor could explain this: _____

Student's Practical Laboratory Experiment

APPLICATION OF CHEMISTRY SKILLS

OBJECTIVE

The objective of this experiment is to use chemistry concepts and tools in a scientific laboratory to determine the chemical properties of unknown pure solid as well as pure liquid. You will work on the questions: *"What is this cubic block made up of?"* and *"Which compound is this liquid made up of?."* To answer these questions, you may need to determine intrinsic properties of your samples such as density. Intrinsic properties are those which do not change when the amount is changed. You will need to compare physical properties of your samples to those of similar known materials found in the literature. This way you might find a perfect match and then know the chemical identity—whether it is an element or compound. You will find, as you study your cubic block and unknown liquid and characterize them, that you need to use some laboratory skills and scientific knowledge you learned before. You will thus review some introductory chemistry concepts and skills.

LEARNING OUTCOMES

After this experiment, students will be able to:

1. Determine the density of a solid and unknown liquid.
2. Determine the chemical identity of unknown pure solid and liquid.
3. Support theoretical information with experimental evidence.

SAFETY NOTES

Handle glassware with care and inspect it for cracks and sharp edges before use. Work in a fume hood or close to an air vent to avoid inhaling fumes from the liquids. Immediately wash off the chemicals should you make skin-contact. Always wear safety goggles and shoes when doing experiments. *When finished, wash your hands with soap and water.*

PRE-LAB ASSIGNMENT

Physical properties are those displayed by the object with its particles chemically unchanged. *Chemical properties* are those displayed by the object as its particles change into different species.

Classify the properties in this table as either Physical or Chemical:

	Physical/ Chemical?		Physical/ Chemical?		Physical/ Chemical?
Melting point		Conductance (thermal/electrical/ magnetic)		Reducing ability	
Boiling point		Acidity		Stability	
Density		Physical state (solid/liquid/gas)		Oxidizing ability	
Solubility		Malleability		Explosiveness	
Odor		Color		Corrosive	
Texture		Inertness		Combustible	
volatility		Flammable		Basicity	
Shape		Reactivity		Causticity	

Materials Needed:

1. About 50 mL of unknown liquid—one of different pure liquids
2. A 20 mL pipet
3. A 125 mL Erlenmeyer flask
4. A less than 1 cm width, cubic block/rectangular prism—one of different pure metal blocks
5. A 100 mL graduated cylinder
6. A 10 cm ruler
7. A calculator
8. An electronic mass-balance

EXPERIMENTAL PROCEDURE

Part I

1. Determine *the density of your solid* cubic block:

   a) Weigh your cubic block on the electronic balance: _____

   b) Measure the dimensions of your cubic block using a ruler. Include estimated digit.

   Length _____ Width _____ Height _____

   c) Find the volume of your cubic block:

      i. *Method 1*—by calculation using formula of a rectangular prism *[show set-up]*:

         Calculated volume of the cubic block. *[Check Sig. Figs.]* _____

      ii. Method 2—by water-displacement technique:

         Step 1. Volume of water _____

         Step 2. Volume of water + dry cube _____

         Volume of the cube (step 2 − step 1): _____

   d) Which of the two methods of finding volumes will give you the most accurate density calculation? _____ Explain: _____
   _____
   _____

   e) Use the volume measurement with the most significant figures to calculate the density of your cube. Show calculations and report appropriate *sig. figs.* and *units*.

Density: _____

Instructor's **OK:** _____

Identities of Chemicals 35

Data Analysis

2. Use the following density table of metals to determine which material you have.

Solid	Density (g/cm³)	Appearance and Stiffness
Aluminum	2.70	Silver-gray luster, low Stiffness
Copper	8.96	Red-orange luster, low Stiffness
Iron	7.87	Grayish luster, high stiffness
Nickel	8.90	Light-silver luster, moderate stiffness
Titanium	4.51	Silver-gray, high stiffness
Zinc	7.14	Gray, low stiffness
Cubic Block No. _____		

a) What material is your cubic block? _____.

b) What is the atomic number of the element your cube is made of? _____

c) Given this number, how many protons does this element have? _____

d) How many valence electrons does this element have? _____

e) What is/are the charge(s) of the ion(s) that this metal tends to make? If more than one, list all. _____ ; _____

f) If this metal reacted with oxygen (a redox reaction) to form an ionic compound, what would be the formula of the resulting compound(s)?

   _____ or _____

g) What is/are the name(s) of the ionic compound(s)?

   _____ or _____

Part II

Determine *the density of unknown liquid*.

a) Weigh a clean, empty 125 mL Erlenmeyer flask: _____

b) Measure 20.00 mL of your unknown liquid using a 20 mL pipet. Follow the illustration and steps below:

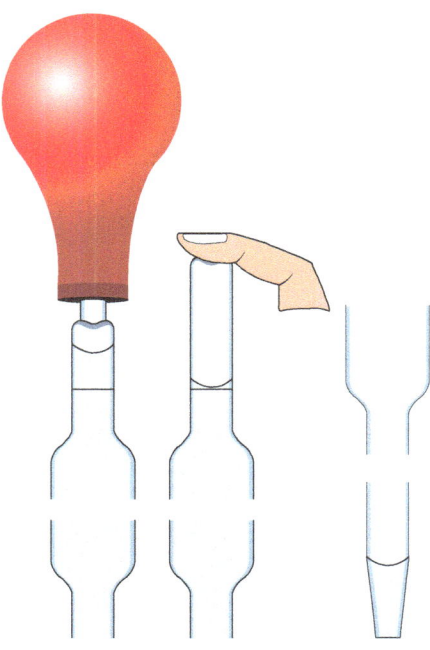

How to use a pipet.

- Clean the pipet with water and detergent, then rinse with plenty of distilled water.
- Rinse once with a small amount of your unknown liquid, and then draw liquid past the fill mark.
- Quickly replace the filler bulb with your finger and gently break seal to allow excess liquid to run out into a waste container until the bottom of the meniscus is *exactly* on the fill mark.
- Lift your finger to let the liquid drain from the pipet into your Erlenmeyer flask.
- *Do not* blow or shake the last bit of liquid at the tip of the pipet into the flask!

c) Re-weigh the flask containing your unknown liquid and record the mass exactly as displayed: _____

d) Calculate the mass of your unknown liquid [Entry (c) − (a)]:

e) Calculate the density (D) of your liquid $\left[D = m/v\right]$

f) Match your calculated density to the density values (at 25°C) in the table to identify your unknown liquid:

Liquid	Density (g/mL)	Appearance, Odor, & Water Solubility
Hexane (C_6H_{14})	0.655	Colorless, slight odor, insoluble
Diethylether [$(C_2H_5)_2O$]	0.713	Colorless, slight odor, insoluble
Toluene ($C_6H_5CH_3$)	0.867	Colorless, strong odor, insoluble
Dimethylformamide [$(CH_3)_2NC(O)H$]	0.944	Colorless, strong odor, soluble
Dimethylsulfoxide [$(CH_3)_2SO$]	1.092	Colorless, odorless, soluble
Dichloromethane (CH_2Cl_2)	1.327	Colorless, sweet odor, insoluble
Chloroform ($CHCl_3$)	1.498	Colorless, strong odor, insoluble
Unknown Liquid No. _____		

- Write the chemical formula of your compound _____.
- All these compounds contain carbon, C. In which Group No. is carbon? _____
- How many and what type of bonds does carbon form? _____, _____
- What type of compounds are these liquids? _____

Instructor's **OK:** _____

END OF LAB REPORT

THESE ARE MY OWN NOTES
(from listening to lectures, watching YouTube videos, etc.)

Date: _____

Date: _____

Date: _____

Date: _____

I wish the professor could explain this: _____

Identities of Chemicals 43

QUANTITIES OF CHEMICALS 3

Compound Stoichiometry

The term *stoichiometry* comes from the Greek and English words *"stoikheion"* for element and *"-metry"* for quantity, so that stoichiometry refers to "element quantity." More precisely, stoichiometry *means relative quantities* and can be applied to elements, compounds, and other things related mathematically to the elements. Remember that *chemical elements* are really different types of atoms and, in practice, we are not able to work with individual atoms or even a few atoms. We use the atomic masses shown in the Periodic table as molar masses (in g/mol) of each element. Having grams per mole (g/mol) for each element allows us to <u>convert grams into *"moles"*</u> abbreviated "mol" of different elements and *vice versa*. Moles are sufficient for most stoichiometry calculations. But if we want to <u>*find the actual number of particles*</u> (atoms, ions, or molecules, formula units) we can use Avogadro's number, N_A, which is 6.022×10^{23} particles/mol. As an example, let us *calculate the number of H_2O molecules in 4.00 g of liquid water*. The first thing we always do is write down the quantity we want to convert. Then we multiply it by appropriate conversion factors until we have the units we want.

$$4.00 \, g \, H_2O \times \frac{1 \, mol \, H_2O}{18.02 \, g \, H_2O} \times \frac{6.022 \times 10^{23} \, H_2O \, molecules}{1 \, mol \, H_2O}$$

$$= 1.34 \times 10^{23} \, H_2O \, molecules$$

Practice Questions

3.1. How many Fe atoms are present in 1.5 g of iron, Fe?

3.2. How many CO_2 molecules are present in 2.94 g CO_2?

3.3. Calculate the number of moles in 8.02×10^{20} Fe atoms

My Professor Says:

Notice that the final answer must have the same uncertainty as the quantity being converted—the same number of *significant figures* in this case since we are multiplying and dividing. If we were adding measured quantities, then the final answer would have to have the same number of *decimals* as the quantity converted. For liquids though, volume is more convenient to use than mass. So, the problem we just discussed could be stated "*calculate the number of H_2O molecules in 4.00 mL of water with density 0.9985 g/mL.*" We would start with 4.00 mL:

$$4.00 \;\cancel{mL\, H_2O} \times \frac{0.9985 \;\cancel{g\, H_2O}}{1 \;\cancel{mL\, H_2O}} \times \frac{1 \;\cancel{mol\, H_2O}}{18.02 \;\cancel{g\, H_2O}} \times \frac{6.022 \times 10^{23} \; H_2O \; molecules}{1 \;\cancel{mol\, H_2O}}$$

$$= 1.33 \times 10^{23} \; H_2O \; molecules$$

Just like grams, "moles" tell us the quantity, even if we do not measure moles directly. If we have *2 mol* of sodium phosphate, *2(Na_3PO_4)* formula units, we can be asked, "How many moles of each element does it have?." The answer *is 2(Na_3PO_4) = 2 × 3 mol Na, 2 × 1 mol P,* and *2 × 4 mol O* or two times each subscript in the formula. Another question can be "*What is* the mass contribution of each element, and *the* final *molar mass of the compound*, Na_3PO_4?." To answer this question, we need to look up the *molar mass* of each element in the periodic table on page 23. The mass contributions are:

3 **Na** atoms × 22.99 g/mol	=	68.97 g/mol **Na**
1 **P** atoms × 30.97 g/mol	=	30.97 g/mol **P**
4 **O** atoms × 16.00 g/mol	=	64.00 g/mol **O**
Compound molar mass	=	**163.94 g/mol** Na_3PO_4

Do I get it? Let me check:

Practice Questions

3.4. How many moles of oxygen are present in 1.0 mol of aluminum oxide, $Al_2(SO_4)_3$?

3.5. Calculate the molar mass of $Al_2(SO_4)_3$.

My Professor Says:

Do I get it? Let me check:

We can also be asked, "What are the *percentages of each element in one mole of the compound*, Na$_3$PO$_4$?." To get percentage we multiply the fraction by 100. That is amount of a component divided the by the sum of components, multiplied by 100. In this case, amount has units of g/mol. So, the percentages are:

$$\frac{68.97\,g/mol\,Na}{163.94\,g/mol} \times 100 = 42.07\%\,\boldsymbol{Na}$$

$$\frac{30.97\,g/mol\,Na}{163.94\,g/mol} \times 100 = 18.89\%\,\boldsymbol{Na}$$

$$\frac{64.00\,g/mol\,Na}{163.94\,g/mol} \times 100 = 39.04\%\,\boldsymbol{Na}$$

$$= \mathbf{100.0\%}$$

Now, if we have an unknown sample analyzed in the laboratory, and it is found to contain 42.07% **Na**, 18.89% **P**, and 39.04% **O** by mass, can we <u>derive the empirical formula of the compound</u>, that is, a formula with the simplest ratio of elements? Let us see.

Since *percent* means out of 100, we can assume that we have 100 g of sample. This now means we have 42.07 g **Na**, 18.89 g **P**, and 39.04 g **O**. Next, we convert these grams to moles and use whole number moles as subscripts:

$$42.07\,\cancel{g\,Na} \times \frac{1\,mol\,Na}{22.99\,\cancel{g\,Na}} = 1.830\,mol\,\boldsymbol{Na}$$

$$18.89\,\cancel{g\,P} \times \frac{1\,mol\,P}{30.97\,\cancel{g\,P}} = 0.6100\,mol\,\boldsymbol{P}$$

$$39.04\,\cancel{g\,O} \times \frac{1\,mol\,O}{16.00\,\cancel{g\,O}} = 2.440\,mol\,\boldsymbol{O}$$

Practice Questions

3.6. Calculate the percentages of aluminum, Al, S, and oxygen, O, in Al$_2$(SO$_4$)$_3$.

3.7. Calculate the percentages of aluminum, sulfur, and oxygen elements in Al$_2$(SO$_4$)$_3$

My Professor Says:

© Mr. Rashad/Shutterstock.com

The *pseudo-empirical* formula is $Na_{1.830}P_{0.6100}O_{2.440}$.

The easiest way to *convert* these *decimals into whole numbers* is to divide each by the smallest one among them; by 0.6100 in this case. So, the *true empirical formula* is $Na_{3.000}P_{1.000}O_{4.000}$, which is written simply as Na_3PO_4.

Converting a pseudo-empirical formula into true empirical formula may involve a few more math steps in addition to dividing each number of moles by the smallest one. Let us consider an example: "*Elemental analysis of ascorbic acid showed that it contained 40.91% carbon, 4.587% hydrogen, and 54.50% oxygen by mass. Find its empirical formula.*"

Assuming 100.0 g sample, the masses are 40.91 g C, 4.587 g H, and 54.50 g O. Converting these to moles we get:

$$40.91 \, \cancel{g\,C} \times \frac{1\,mol\,C}{12.01\,\cancel{g\,C}} = 3.406\,mol\,C$$

$$4.587 \, \cancel{g\,H} \times \frac{1\,mol\,H}{1.01\,\cancel{g\,H}} = 4.541\,mol\,H$$

$$54.50 \, \cancel{g\,O} \times \frac{1\,mol\,O}{16.00\,\cancel{g\,O}} = 3.406\,mol\,O$$

The *pseudo-empirical* formula is $C_{3.406}H_{4.541}O_{3.406}$.

Dividing by the smallest moles, by 3.406 in this case, we get $C_{1.000}H_{1.333}O_{1.000}$.

Note that for these calculations, the decimal number must be *within ± 0.05* of the whole number to be rounded into that whole number. If it is not, just as 1.333 is

Do I get it? Let me check:

© AVIcon/Shutterstock.com

Practice Questions

3.8. The herbicide, dixanthogen, consists of 29.73% **C**, 4.16% **H**, 13.20% **O**, and 52.91% **S**. What is its *empirical formula*?

3.9. Find the empirical formula of glyoxylic acid which contains 32.5% C, 2.70% H, and 64.8% O.

My Professor Says:

not, we *convert* that *decimal number into a common fraction* as follows:

Step 1. Write down the decimal divided by 1.

Step 2. (a) For *recurring* decimals, multiply both top and bottom by the *repeating* digit.

(b) For *nonrecurring* decimals, multiply top and bottom by *10 for every decimal,* that is, *by 100 for two decimals, by 1000 for three, etc.*

Step 3. Simplify the fraction into a common fraction of integers.

Now, the empirical formula for ascorbic acid with moles *as integers or common fractions* of integers will be **$C_1 H_{4/3} O_1$**.

The next step is to clear the fraction by multiplying each number by the denominator (3 in this case), which gives us **$C_3 H_4 O_3$** as the final empirical formula for ascorbic acid. While formula units of ionic compounds always have the simplest ratio of moles, that is, simplified subscripts, very few molecular compounds occur in simplest ratios of moles in nature. The table below shows some examples:

Molecular Compound	Molecular Formula	Empirical Formula	Conversion Factor
Water	H_2O	H_2O	1
Ammonia	NH_3	NH_3	1
Tetraphosphorous decaoxide	P_4O_{10}	P_2O_5	2
Benzene	C_6H_6	CH	6
Glucose	$C_6H_{12}O_6$	CH_2O	6
Ascorbic acid	$C_6H_8O_6$	$C_3H_4O_3$	2

Do I get it? Let me check:

Practice Questions

3.10. What is the empirical formula of ethylene glycol, $C_2H_6O_2$?

3.11. What is the empirical formula of ribose, $C_5H_{10}O_5$?

3.12. Molecular formula is a whole-number multiple of empirical formula. Find the molecular formula of butyric acid if its molar mass = 88.11 g/mol, and its empirical formula is $C_2H_4O_1$.

My Professor Says:

In the laboratory, scientists have an instrument called a mass spectrometer which can measure the molar mass of molecules. In the table above, you may have realized that a factor is used to convert (by dividing) subscripts from molecular formula to empirical formula. So, if we know molar mass of the molecular compound as it exists in nature, as well as its empirical formula, we can find the conversion factor. To do the opposite, that is, convert the empirical formula into the molecular formula we will need to multiply subscripts of empirical formula with the factor. Given the molar mass of ascorbic acid of 176.14 g/mol, and knowing its empirical formula, $C_3H_4O_3$, we can determine its molecular formula.

All we need is the conversion factor.

$$Conv.\,Factor = \frac{Molecular\,Formula\,mass}{Empirical\,Formula\,mass}$$

$$= \frac{176.14\,g/mol}{[3(12.01) + 4(1.01) + 3(16.00)]\,g/mol} = 2$$

Now, the molecular formula of ascorbic acid should be $[C_3H_4O_3]_{\times 2} = C_6H_8O_6$.

Solution Stoichiometry

A solution is a homogeneous mixture of the solvent and the solute, that is, the chemical substance that is dissolved. In solution stoichiometry we are interested in *concentration*, that is, how much chemical species (in moles) are present in solution (as liters, L). Since most soluble chemical substances are compounds and

Do I get it? Let me check:

Practice Questions

3.13. Calculate the molarity of a 100 mL solution that was made by dissolving 15.0 g of NaOH in distilled water.

3.14. Exactly 5.00 mL of 1.18M KCl solution was diluted to 50.0 mL. What is the molarity of the final, dilute solution?

My Professor Says:

© Mr. Rashad/Shutterstock.com

the mass (in g) dissolved is usually known, we simply convert that mass into moles and divide by the final volume (in L) of solution. When concentration is expressed this way (as mol/L), it is called *molarity*. Note that the units of molarity are mol/L but sometimes "*M*" is used as units, for reading or labeling convenience.

To *determine grams of a compound* that is, or must be, weighed out and dissolved to make a specific volume (mL) of a given molarity, we build a dimensional analysis set-up, starting with the volume (mL). We multiply it by 1 L over 1000 mL so that "mL" can cancel out. Then, given molarity, we multiply with moles as numerator and 1 L as denominator so "L" can cancel out. Finally, we multiply with molar mass (g/mol) of the compound so "mol" can cancel out, leaving grams as final units.

To *find final molarity* after a solution of known molarity was diluted to a known final volume, use the dilution equation ($M_{initial}V_{initial} = M_{final}V_{final}$). You can use the same dilution equation to find initial molarity. Here is a problem that we can use as example: "*What mass (in g) of $KCl_{(s)}$ must be dissolved to make 25.0 mL of a 1.18 M KCl solution?*"

$$25.0 \; ml \; Soln \times \frac{1 \; L \; Soln}{1000 \; ml \; Soln} \times \frac{1.18 \; mol \; KCl}{1 \; L \; Soln} \times \frac{74.55 \; g \; KCl}{1 \; mol \; KCl}$$

$$= 2.20 \; g \; KCl$$

Concentration can be expressed in various other ways in addition to molarity. It can be expressed as molality (*m*), mole

Do I get it? Let me check:

© AVIcon/Shutterstock.com

My Professor Says:

fraction (χ), parts by mass (% $^m/_m$; ppm, ppb, etc.), or parts by volume (%$^{vol}/_{vol}$). Parts by mass is typically used when the solute is a solid and parts by volume is used when the solute is a liquid.

We already learned that *molarity* is defined as *moles of solute divided by liters of solution* and has units of mol/L. *Molality* is defined as *moles of solute divided by kilograms of solvent* and has units of mol/Kg (or molal, *m*). *Mole fraction* is defined as *moles of one component of interest divided by the sum of moles of all components present in the solution*. *Parts by mass* is defined as grams of solute divided by grams of solution (i.e., solvent + solute) and when this ratio is multiplied by 100, we get % $^m/_m$. When it is multiplied by 1 000 000, we get parts per million (ppm), and when it is multiplied by 1 000 000 000, we get parts per billion (ppb), and so on.

Inventory of Solutions

Inventory is a list of the major solute species that are present in a solution in relatively high concentrations. In order to list the individual species that are present in aqueous solution, you must remember that acids and ionic compounds break up into two different ionic species (cation and anion) when they dissolve. In contrast, molecular compounds simply get dispersed and individual molecules do not break up.

While *ionic compounds produce ions according their ratio* in the formula when they break up, *acids always break up into 1:1 ratio*

Do I get it? Let me check:

Practice Questions

3.15. A student dissolved 20.4 g $CaBr_2$ in exactly 150.0 g of deionized water. She determined the density of this solution to be 1.045 g/mL. What is the concentration of her solution, expressed as …?

a) Molarity (*M*)

b) Molality (*m*)

c) Mole fraction (χ) of $CaBr_2$

My Professor Says:

of hydrogen ion (H^+) and the anion. The H^+ of the acid always attaches itself to H_2O (from the solvent) and appears as hydronium ion (H_3O^+). Even acids which have two or more acidic (or dissociable) hydrogens lose only one H^+ and the rest of the hydrogens remain part of the anion.

There are two types of acids—*strong acids* (which we will call the big six; namely HCl, HBr, HI, HNO_3, $HClO_4$, and H_2SO_4) and *weak acids*. "Strong" acid means every single molecule of that acid breaks up into hydrogen ion (H^+) and the anion. On the other hand, "weak" acid, which is any acid other than the big six, means only a few molecules (less than 10%) of that acid break up into H^+ and the anion.

What we just said about inventory of acid solutions lead to the following general statements:

- Inventory of *a strong acid solution* is simply hydronium (H_3O^+) ion and the anion, each at the original concentration of the strong acid.

- Inventory of *a weak acid solution* is the weak acid itself at its original concentration. The hydronium (H_3O^+) ions and the anion are negligible.

What about bases? *What is a base?* A base is any chemical species that accepts H^+ from an acid in solution. If there are no acid molecules in solution, a base will remove H^+ from H_2O, attach that H^+ to itself to form an acid, and by doing so generate OH^- ions. *Anions of weak acids are strong bases.* The weaker the acid, the stronger the base its anion will be. The bases are typically

Do I get it? Let me check:

Practice Questions

3.16. Write the inventory for each of the following aqueous solutions. Show the expected concentration of each species:

a) 0.20 **M** $Ca(NO_3)_{2\,(aq)}$

_____

b) 0.20 **M** $HF_{(aq)}$

_____

c) 0.20 **M** $HCl_{(aq)}$

_____

d) 0.20 **M** $H_2SO_{4\,(aq)}$

_____

e) 0.20 **M** $HCO_3^-{}_{(aq)}$

_____

My Professor Says:

made available as ionic compounds. For example, NaHCO$_3$ is a base because when it dissolves HCO$_3^-$ is set free. Every HCO$_3^-$ released instantly grabs H$^+$ from H$_2$O to form H$_2$CO$_3$, its conjugate acid, leaving OH$^-$ in solution. These observations about inventory of base solutions lead to the following general statements:

- Inventory of a strong base is simply hydroxide (OH$^-$) ions and the acid formed by the base, each at the original concentration of the strong base, that is, anion from the ionic compound.

- Inventory of *a weak base (i.e., anions of strong acids)* is the weak base itself at its original concentration. These do not remove/accept H$^+$ from water because strong acids do not exist as molecules in solution.

One exception, in that it is not an anion of an ionic compound, is ammonia, NH$_3$, which to a small extent accepts H$^+$ from H$_2$O and therefore is a weak base. Since a negligible number of weak acid species release H$^+$ ions, and negligible weak base species accept H$^+$ ions, when a given volume of a weak acid (WA) and volume of a weak base (WB) are mixed, we can use the dilution equation to determine the molarity of the weak acid or the weak base in the final solution:

$$M_{(WA,initial)} \times V_{(WA,initial)} = M_{(WA,final)} \times V_{(final)}$$
or $M_{(WB,initial)} \times V_{(WB,initial)} = M_{(WB,final)} \times V_{(final)}$.

Remember that in these cases, the final volume is the sum of initial volumes of weak acid and weak base.

Do I get it? Let me check:

Practice Questions

3.17. Give a list of chemical species that are present in 60.0 mL of 0.10 **M** HCN$_{(aq)}$ mixed with 50.0 mL of 0.10 **M** NH$_{3(aq)}$

THESE ARE MY OWN NOTES
(from listening to lectures, watching YouTube videos, etc.)

Date: _____

Date: _____

Date: _____

Date: _____

I wish the professor could explain this: _____

Student's Practical Laboratory Experiment

STOICHIOMETRY OF CHEMICAL COMPOUNDS

OBJECTIVE

The objective of this experiment is to determine the empirical formula of ionic hydrate from mass measurements.

Many salts, that is, ionic compounds that are crystallized from aqueous solutions are found to be weakly coordinate with several water molecules within the crystal lattice. Water fits in at specific sites within the lattice and is therefore present in stoichiometric amounts. These types of compounds are called *ionic hydrates*. Some examples are $FeCl_3 \cdot 6H_2O$, $CuSO_4 \cdot 5H_2O$, and so on.

In this experiment, a weighed sample of magnesium sulfate hydrate will be heated to drive off all the water, and then reweighed to determine the mass of the anhydrous residue and that of the water driven off:

$$MgSO_4 \cdot nH_2O(s) \rightarrow MgSO_4(s) + nH_2O(g).$$

The number n is an integral value such as 1, 2, 3, and so on. The masses of anhydrous magnesium sulfate and water are each converted to moles, and the calculated ratio of *moles of water* (driven off) to *moles of anhydrous magnesium sulfate* (remaining) is equal to n.

LEARNING OUTCOMES

After this experiment, students will be able to:

1. Produce an anhydrous salt by heating an ionic hydrate to a constant mass
2. Determine the amount, in moles, of water of hydration in salt crystals
3. Determine the empirical formula of an ionic hydrate

SAFETY NOTES

Do not leave the open flame unattended. Shut the Bunsen burner off immediately after use by closing the gas valve. Handle the hot crucible with tongs and place it on a clay tile, rather than directly on the countertop. Always wear safety goggles and shoes when doing experiments. *When finished, wash your hands with soap and water.*

Quantities of Chemicals **57**

PRE-LAB ASSIGNMENT

Write down what you think about the following:

1. How will you determine when all the water has evaporated?

   _____
   _____
   _____

2. What problems can over-heating the hydrate cause?

   _____
   _____
   _____

3. List the possible sources of error in this experiment and how you would avoid them.

   _____
   _____
   _____

Materials Needed:

1. Some magnesium sulfate hydrate ($MgSO_4 \cdot nH_2O$) crystals
2. A Bunsen burner, porcelain crucible, tripod, clay triangle, crucible tongs, striker/lighter
3. A calculator

EXPERIMENTAL PROCEDURE

1. Assemble the tripod and burner as shown below. Place the clay triangle to support the crucible. Adjust the air intake of the burner to eliminate orange color in flame, which may deposit soot on the crucible.

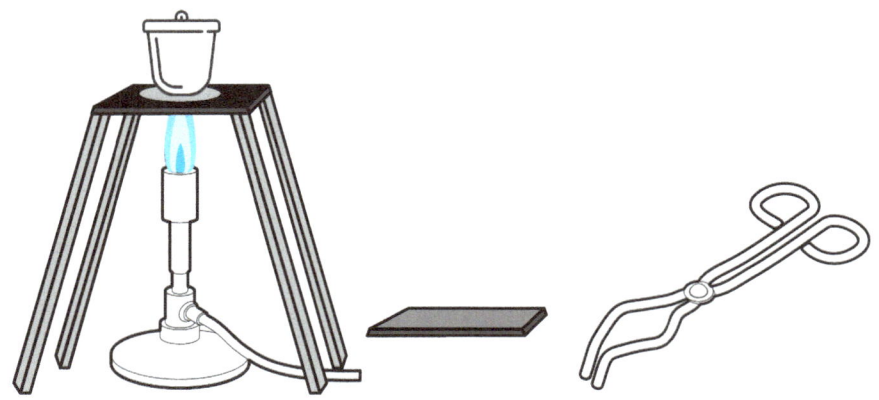

Images modified from © Rvector/Shutterstock.com and © NinjaStudio/Shutterstock.com

2. Place clean porcelain crucible on the clay triangle and tripod. Heat it by a direct flame for 5 minutes to drive off any moisture. Remove the crucible with tongs and put it on the tile to cool to room temperature (about 10 minutes). Weigh it with its cover to the nearest 0.001 g as shown above.

3. Place about 1.5 g of solid magnesium sulfate hydrate into the crucible and weigh the crucible/cover and contents together to the nearest 0.001 g.

4. Return the crucible to the clay triangle, using the tongs. Leave the lid slightly open and heat the crucible for about 15 minutes.

5. Move the burner and allow the crucible to cool to room temperature on a wire gauze pad, and again weigh the covered crucible to the nearest 0.001 g.

6. To ensure complete reaction, repeat the heating and weighing process until two successive readings agree to within 0.005 g. Use the last weight recorded as your final data.

Record all data below with appropriate number of significant figures and units.

1. Mass of crucible (*cleaned by heating*) _____
2. Mass of crucible + $MgSO_4 \cdot nH_2O$ (*the hydrate*) _____
3. Calculated mass of hydrate [step 2 − step 1]: _____

HEAT CRUCIBLE + CONTENTS, COOL & WEIGH

4. Mass of crucible + *anhydrous* $MgSO_4$ (after first heating): _____

HEAT CRUCIBLE + CONTENTS, COOL & WEIGH

5. Mass after heating second time: _____

HEAT CRUCIBLE + CONTENTS, COOL & WEIGH

[IF step 5 − step 4 < 0.005 g do not heat again. IF step 5 − step 4 > 0.005 g heat again]

6. **Mass** after heating third time (*only if needed*): _____

7. **Mass** of anhydrous magnesium sulfate, $MgSO_4$ *[step 6 or 5 − step 1]*: _____

8. **Moles** of anhydrous magnesium sulfate:

    a) *Find the molar masses (g/mol) of the elements in $MgSO_4$, from the periodic table, and add them up to get molar mass of $MgSO_4$:* _____

    b) *Now use the molar mass of $MgSO_4$ in a dimensional analysis set-up to convert grams $MgSO_4$ in step 7 into moles $MgSO_4$:*

    moles $MgSO_4$: _____

9. **Mass** of water, H_2O, eliminated *[step 3 − step 7]*: _____

10. **Moles** of anhydrous magnesium sulfate:

    a) *Find the molar masses (g/mol) of the elements in H_2O, from the periodic table, and add them up to get molar mass of H_2O:* _____

    b) *Now use the molar mass of H_2O in a dimensional analysis set-up to convert grams H_2O in step 7 into moles H_2O:*

    moles H_2O: _____

11. **The ratio of moles H_2O to $MgSO_4$ in the hydrate:**

    a) *Divide moles H_2O by moles $MgSO_4$ [step 10b ÷ step 8b]* _____

    b) *Round to a whole number:* _____

    [This is the value of "n" in the hydrate formula]

12. Write the chemical formula of the *ionic hydrate*: _____

POST-LAB QUESTIONS

1. Write your compound formula again here: _____

2. What is the ratio of **Mg^{2+}** to **SO$_4^{2-}$** ions? _____

3. What is the ratio of water molecules **H$_2$O** to **MgSO$_4$**? _____

4. What is the ratio of **Mg^{2+}** to **O** atoms? (incl. O's in water) _____

5. Calculate the *molar mass* of this entire compound? (use 4 *sig. figs.* from periodic table)

6. Calculate the *element composition* (% of **Mg**, **S**, **O**, and **H**) in this compound (3 *sig. figs.*):

    a) **Mg** _____

    b) **S** _____

    c) **O** _____

    d) **H** _____

7. What is the % *by mass* of H$_2$O in this compound? (3 *sig. figs*)

8. If you have 24.477 g of this compound, how many **O** *atoms* are in it?

9. What is the empirical formula of a compound that is composed of 6.00 g of carbon, and 2.01 g of hydrogen? *[Convert grams to moles of each atom, then into whole number ratios]*

–o0o–   END OF LAB REPORT

THESE ARE MY OWN NOTES
(from listening to lectures, watching YouTube videos, etc.)

Date: _____

Date: _____

Date: _____

Date: _____

I wish the professor could explain this: _____

Quantities of Chemicals

Quantities of Chemicals

CHEMICAL EQUATIONS

Molecular (or Reagent) Equations

If chemistry is a game, then chemical species are the players. In chemistry, we focus on the changes that chemical species undergo as they interact with each other. Different compounds may interact with each other or with individual atoms or ions. Their interactions are called chemical reactions and we represent them as *balanced chemical equations*. We will discuss three types of chemical reactions in this course—precipitation, reduction–oxidation (redox), and acid–base reactions.

A molecular equation uses formulas to express the identities and quantities (as coefficients) of substances involved.

Practice Questions

4.1. *Beryllium* and *silver nitride* react to produce *beryllium nitride* and *silver*. Circle the correct chemical equation:

a) $3Br + Si_3N_2 \rightarrow Br_3N_2 + 3Si$

b) $2B + AgN_3 \rightarrow B_2N_3 + Ag$

c) $Be + 2SrN \rightarrow BeN_2 + 2Sr$

d) $3Be + 2Ag_3N \rightarrow Be_3N_2 + 6Ag$

e) $Be + 2AuN \rightarrow BeN_2 + 2Au$

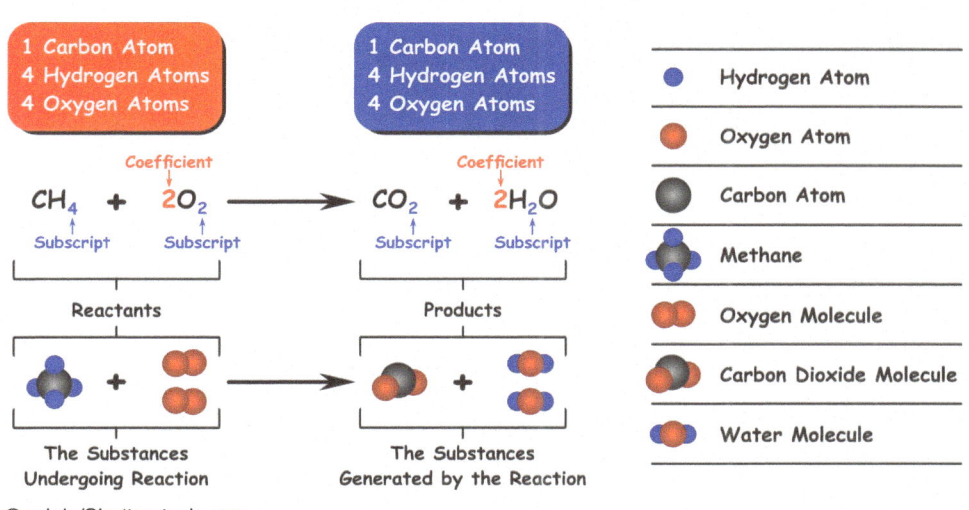

© udaix/Shutterstock.com

My Professor Says:

Coefficients are used to make sure the number of atoms present before a reaction are all there after the reaction. We say the equation is balanced when correct coefficients are there.

To study a chemical reaction in the laboratory, we weigh grams of at least one of the chemical species on the left side of the yield arrow, called reactants. Reactants interact best when they are dissolved. So, pre-made solutions of known concentrations called reagents are usually kept in the laboratory and used for specific reactions. Species on the right side of the yield arrow (called products) are not there before the reaction takes place. After the species interact, that is, bond with each other, exchange partners, or transfer electrons, they become different species called products. One species can also change, by separating into smaller species also called products. By balancing the equation (which we will discuss later) we simply make sure that the number of atoms before the reaction is the same after the reaction.

Do I get it? Let me check:

Practice Questions

4.2. *Ammonium chloride* and *sodium nitrite* reacts in aqueous solution to produce *nitrogen gas*, *sodium chloride*, and *water* Write a balanced chemical equation.

My Professor Says:

Do I get it? Let me check:

Total and Net Ionic Equations

We often write the net chemical equation instead of molecular equation, where only the species that undergo a chemical change are shown, as follows:

1. *Start by writing the total ionic equation,* where you separate every aqueous compound in the reagent equation into cation and anion. Note that when separating cation and anion, the subscript of the cation becomes the superscript of the anion and *vice versa*. Compounds that appear as liquid, gas, or solid in the chemical equation should NOT be separated.

2. *Now write a net ionic equation,* where you remove cations and anions that appear unchanged on both sides of the yield arrow. For a *precipitation reaction*, one product is always a solid ionic compound.

Reaction Stoichiometry

Let us now learn how to calculate the amount of a product that can form using the amount of reactant weighed to start the reaction with, and a balanced equation. The first thing we do is write down the *reactant mass (in g)* and *multiply it by* its molar mass (*as 1 mol/___ g*, so grams can be canceled) to convert it into moles, *followed by moles of product over moles of the reactant* (shown as coefficients in the balanced equation).

Practice Questions

4.3. For each reagent equation write the total ionic equation and the net ionic equation

a) $MgCO_3(aq) + 2HCl(aq) \rightarrow MgCl_2(aq) + H_2O(l) + 2CO_2(g)$

_____

_____

_____

b) $Na_2CO_3(aq) + Ni(NO_3)_2(aq) \rightarrow 2NaNO_3(aq) + NiCO_3(s)$

_____

_____

_____

c) $2NH_4Br(aq) + Ag_2SO_4(aq) \rightarrow (NH_4)_2SO_4(aq) + 2AgBr(s)$

_____

_____

_____

My Professor Says:

Canceling moles of reactants gives us moles of the product. This is how we switch to moles of the product. We *finish the set-up by multiplying with the molar mass of the product* (as ____ g/mol) to convert product moles to grams. Using 4.00 g CH_4 to start the reaction we just showed [$CH_4(g) + 2O_2(g) \rightarrow CO_2(g) + 2H_2O(g)$], how much H_2O (in g) can be formed? Here is how the set-up should look:

$4.00 \, g\, CH_4 \times \dfrac{1 \, mol\, CH_4}{16.04 \, g\, CH_4} \times \dfrac{2 \, mol\, H_2O}{1 \, mol\, CH_4} \times \dfrac{18.02 \, g\, H_2O}{1 \, mol\, H_2O}$
$= 8.99 \, g\, H_2O$

This is called reaction stoichiometry.

The Limiting Reactant

The limiting reactant is the reactant that is available in less amount than required for complete combination with the available amount of the other reactant. *To determine the limiting reactant* use method 1 if the balanced chemical equation has products shown. Use method 2 if products are not shown in the balanced chemical equation.

Method 1:

Build a dimensional analysis set-up to convert grams of the first reactant to grams of the target product. Build a dimensional analysis set-up to convert grams of the second reactant to grams of the same target product. The reactant that gives the lowest amount of product is the limiting reactant.

Do I get it? Let me check:

Practice Question

4.4. Propane reacts with excess oxygen as follows:

$C_3H_8(g) + 5O_2(g) \rightarrow 3CO_2(g) + 4H_2O(g)$

How much propane, C_3H_8, will react with excess oxygen, O_2, to produce 85.2 g CO_2?

What is the first thing we always write down to start a set-up? We write down the quantity given, which is 85.2 g O_2.

STEP 1: Let's do that. STEP 2: We must now convert it to _____ of O_2. This is what we do in chemistry. We convert grams to moles. STEP 3: Switch to moles of what we want –CO_2 here– and convert them to grams. Let me try:

My Professor Says:

Do I get it? Let me check:

Method 2:

Circle the given grams of the first reactant and label it "AVAILABLE." Build a dimensional analysis set-up to convert grams of the second reactant to grams of the first reactant that are "REQUIRED" for complete combination. Example: In the reaction $2SO_2(g)[3.00g] + O_2(g)[2.00g] \rightarrow$ **products**, conversion of 2.00 g $O_2(g)$ ends with 8.01 g $SO_2(g)$ required. Since only 3.00 g is available, $SO_2(g)$ will limit the reaction.

- If grams "REQUIRED" is greater than grams "AVAILABLE" then the first reactant is the limiting reactant.

- If grams "REQUIRED" is less than grams "AVAILABLE" then the first reactant is in excess and the second one is the limiting.

Balancing Chemical Equations

Remember that:

- A chemical equation is a statement. We should use correct chemical symbols and correct chemical formulas for the chemical species as they are found in the laboratory.

- Every element in the equation, whether it is an ion or atom, as well as overall charge must be balanced, that is, quantities on the left must be equal to quantities on the right.

Practice Questions

4.5. The reaction of phosphorus and chlorine is controlled to avoid industrial pollution:

$$P_{4(g)} + 4\,Cl_{2(g)} \rightarrow \textbf{products.}$$

If exactly 4.00 g P_4 and 8.00 g Cl_2 are available to react, which reactant will limit the reaction? Show calculations and state your conclusion.

My Professor Says:

Each time you *balance* a nonredox equation follow the steps below:

1. Place a question mark (or zero) in front of each species that is not yet balanced in the equation.

2. You will replace these question marks (or zeros), one at a time, with appropriate coefficients as you balance each element or charge.

3. Write a list of the items (elements as well as charge) that are present in the equation.

4. Mark the items on the list that appear once on the left and once on the right.

5. Start by writing a coefficient of 1 on the first species and balance each element it contains that is on the right

6. Balance each of the marked items before those that appear more than once on either side.

7. You may use a fractional coefficient to balance an item but then convert the fraction to an integer by multiplying each coefficient already placed, by the denominator.

Do I get it? Let me check:

Practice Questions

4.6. Balance the following chemical equations

a) ___$C_6H_5NO_2$ + ___O_2 → ___CO_2 + ___H_2O + ___N_2

b) ___$CO(NH_2)_2$ + ___NO + ___O_2 → ___CO_2 + ___H_2O + ___N_2

c) ___C_4H_9NO + ___N_2O + ___OH^- → ___CO_3^{2-} + ___H_2O + ___N_2

d) ___$C_2H_5NO_2$ + ___N_2O + ___OH^- → ___CO_3^{2-} + ___H_2O + ___N_2

My Professor Says:

Precipitation Reactions

A precipitation reaction occurs when two dissolved ionic compounds interact and a solid product identified by "(s)" is formed. Precipitation reactions can be written in one of three ways that we have already discussed—as molecular, total ionic, or net ionic equation. There are some general rules which we can use to predict whether a precipitate, that is, an insoluble compound will be formed when the ions exchange partners:

1. Most *nitrate* (NO_3^-) compounds are *soluble*

2. Most compounds of NH_4^+ and those of *Group 1A* metal ions are *soluble*

3. Most *chloride* (Cl^-) compounds are *soluble* except those with Ag^+, Pb^{2+}, or Hg_2^{2+} ions

4. Most *sulfate* (SO_4^{2-}) compounds are *soluble except* those with Ba^{2+}, Pb^{2+}, or Ca^{2+} ions

5. Most *hydroxide* (OH^-) compounds are *insoluble* except with Ba^{2+}, Ca^{2+}, NH_4^+, or *Group 1A* metal ions

6. Most compounds of *sulfide* (S^{2-}), *carbonate* (CO_3^{2-}), and *phosphate* (PO_4^{3-}) are *insoluble*

Do I get it? Let me check:

Practice Question

4.7. Write the products along with its physical state (*s*, *l*, *g*, or *aq*) and balance the equation.

a) $Na_2S(aq) + AgNO_3(aq) \rightarrow$
_____ + _____

b) $BaCl_2(aq) + (NH_4)_2SO_4(aq) \rightarrow$
_____ + _____

c) $MgCO_3(aq) + HCl(aq) \rightarrow$
_____ + _____

THESE ARE MY OWN NOTES
(from listening to lectures, watching YouTube videos, etc.)

Date: _____

Date: _____

Date: _____

Date: _____

I wish the professor could explain this: _____

Student's Practical Laboratory Experiment
STOICHIOMETRY OF CHEMICAL REACTIONS

OBJECTIVE

The objective of this experiment is to demonstrate how mass, hence moles, of a reactant can be used to calculate the mass or moles of a product based on a balanced chemical equation.

In this experiment, we study the reaction stoichiometry, that is, relative amounts, of magnesium metal, $Mg_{(s)}$, and oxygen gas, $O_{2(g)}$, which react when Mg-metal burns in air. When these two elements react, they yield magnesium oxide, $MgO_{(s)}$. In addition to $O_{2(g)}$, however, air contains nitrogen gas, $N_{2(g)}$. Therefore, another reaction also occurs which yields magnesium nitride, $Mg_3N_{2(s)}$. These two reactions occur in competition with each other when Mg-metal burns in air. We will therefore, follow a sequence of steps listed below to convert any magnesium nitride formed into magnesium oxide, such that all magnesium atoms are ultimately converted into magnesium oxide:

- *Step 1:* Heat magnesium metal in air (in a crucible)
- *Step 2.* After cooling down the reaction mixture, we will add some liquid water, $H_2O_{(l)}$, in order to convert $Mg_3N_{2(s)}$ into magnesium hydroxide, $Mg(OH)_{2(s)}$ accompanied by the release of ammonia gas, $NH_{3(g)}$.
- *Step 3:* We will then heat the reaction mixture to decompose $Mg(OH)_{2(s)}$ into $MgO_{(s)}$ and water vapor, $H_2O_{(g)}$.

LEARNING OUTCOMES

After this experiment, students will be able to:

1. Produce a new chemical product directly and *via* a salvage reaction path
2. Determine the amount of product obtained, that is, actual yield versus the amount expected, that is, theoretical yield.

SAFETY NOTES

Do not leave the open flame unattended. Shut the Bunsen burner off immediately after use by closing the gas valve. Handle the hot crucible with tongs and place it on a clay tile, rather than directly on the countertop. Always wear safety goggles and shoes when doing experiments. <u>When finished, wash your hands with soap and water.</u>

PRE-LAB ASSIGNMENT

1. Write *a balanced* or stoichiometric *equation* for the reaction of *magnesium* and *oxygen gas* to yield *magnesium oxide*. Include physical states.

2. Looking at your chemical equation, do you expect the total mass (of crucible and its contents) to *INCREASE*, or to *DECREASE* after the reaction? _____.
Explain.

_____

_____

_____

Materials Needed

1. Some magnesium chips
2. A Bunsen burner, porcelain crucible, tripod, clay triangle, crucible tongs, striker/lighter
3. A calculator

EXPERIMENTAL PROCEDURE

1. Assemble the tripod and burner as shown below. Place the clay triangle to support the crucible. Adjust the air intake of the burner to eliminate orange color in flame, which may deposit soot on the crucible.

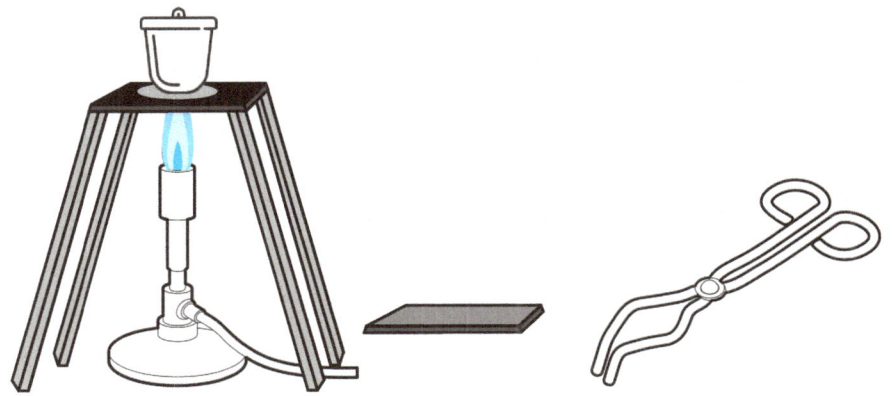

Images modified from © Rvector/Shutterstock.com

2. Heat for five minutes, then remove from the flame, and allow to cool for 5–10 minutes.
3. After cooling, use the crucible tongs to transfer the crucible to the laboratory balance and weigh the crucible. Record this mass to the nearest 0.001 g on the data sheet.
4. Weigh about 0.600 g (± 0.005 g) of magnesium into your crucible. Record the total mass.

Heat Mg metal in air:

5. Return the crucible with the magnesium in it to the clay triangle and adjust the cover on the crucible so that there is a small gap to limit the amount of air entering.
6. Heat the crucible and contents for at least 15 minutes.
7. Carefully remove the hot crucible cover and heat the open crucible for another 10 minutes.
8. Stop heating and allow the crucible and contents to cool to room temperature.

Covert Mg_3N_2 to MgO:

9. Carefully and evenly moisten the cooled residue with 15 drops of deionized water.
10. Return the crucible to the clay triangle and cover as before, leaving a small opening.
11. Heat the crucible for 10 minutes to ensure complete reaction, and complete loss of moisture.
12. Remove cover. Remove the crucible from the flame and allow the contents to cool to room temperature on a clay tile.
13. Weigh the crucible and contents to the nearest 0.001 g. Record this mass on the data sheet.

Data Table: Include units and report the appropriate number of significant figures:

1. Mass of crucible (after first heating) _____
2. Mass of crucible + Mg metal _____
3. Mass of Mg (*Calculation: step 2 − step 1*) _____
4. Matter Observations [*Describe the appearance of the Mg metal before burning*]:
   _____

5. <u>Energy</u> Observations [*Note: Heating only activates Mg atoms so that they can react with O_2 or N_2 molecules. In this case, heat does not drive the reaction. Describe the mixture inside the crucible <u>during the reaction</u>. For example, is it flickering? State whether you think the reaction is endothermic or exothermic. Explain.*]:

   _____
   _____
   _____
   _____

After burning Mg metal in air AND allowing the mixture to cool:

6. Mass of crucible + contents _____
7. Mass of contents only (*Calculation: step 6 − step 1*) _____

After the addition of 15 drops of water, 10 minute re-heating and cooling

8. Mass of crucible + MgO (product) _____
9. Mass of MgO only (Actual yield) _____
10. <u>Matter</u> Observations after you have the final weight. [*Stir the remaining solid and describe the color(s) and texture of the contents. Use these observations and state if you have obtained a pure substance. If not, what could improve your product or the reaction?*]

Pure Mg metal
(silver-grey luster)
© Fablok/Shutterstock.com

Pure MgO
(white)
© SUPIDA KHEMAWAN/Shutterstock.com

Pure Mg_3N_2
(Greenish-yellow)
© Mr. SUTTIPON YAKHAM/Shutterstock.com

_____
_____
_____
_____

POST-LAB QUESTIONS

1. Show your insight in chemistry.

   a) Write a balanced equation for the reaction of magnesium metal and nitrogen gas to yield solid magnesium nitride.

   _____

   b) Write a balanced equation for the reaction of magnesium nitride with water to yield solid magnesium hydroxide and ammonia gas.

   _____

   c) Write a balanced equation for the reaction for thermal decomposition of solid magnesium hydroxide to solid magnesium oxide and water vapor.

   _____

2. Consider the reactants, Mg metal and air. Which one is the limiting reactant?

   _____

3. Calculate the <u>theoretical yield</u> of magnesium oxide, MgO, formed by the mass of magnesium metal ignited in excess air. Show your work including units.

4. Calculate the percent of yield of MgO for your experiment. Show your calculations.

5. Check the box that applies to your results. Give cause and explain. [Was it heating/reactant contact/weighing error/material loss/Impurities/other?]

Check one	% Yield	Why, and what could help?
	Greater than 100%	
	Above 50% but below 100%	
	Below 50%	

–o0o–   **END OF LAB REPORT**

Reduction–Oxidation Reactions

Reduction–Oxidation (also called RedOx) reactions always have one reactant losing electrons (or oxidized) which is called a *reducer* (or reducing agent), and the other reactant gaining electrons (or reduced) which is called an *oxidizer* (or oxidizing agent). Reactants may be simple elements in their natural form, ionic compounds, simple ions, polyatomic ions, or molecular compounds. It is easier to determine the oxidizer and the reducer when reactants are elements, ionic compounds or simple ions since we can compare the charges of a given element before and after the reaction. If charge of a given element becomes more positive, that element has lost electrons (i.e., it was oxidized). If charge of a given element becomes less positive (or more negative), that element has gained electrons (i.e., it was reduced).

While we know the stable charges of many elements and can tell what they are from a compound they have formed, it is not easy to do so when the reactants are polyatomic ions or molecular compounds. These are held together by covalent bonds rather than electrostatic attractions. To determine the oxidizer and reducer in these cases, we *calculate the oxidation number* of a given element in the polyatomic ion or molecular compound before the reaction and after the reaction has occurred.

Practice Questions

4.8. Based on the reaction of potassium and chlorine shown below, which statement is correct?
$2K(s) + Cl_2(g) \rightarrow 2KCl(s)$

a) Chlorine is the reducing agent

b) Potassium is the reducing agent

c) Potassium is the oxidizing agent

d) Neither Cl_2, K is reduced nor oxidized

e) This is a precipitation reaction

4.9. Complete the table for each highlighted element:

	Charge (Ox. No.)	Oxidized/ reduced?
a) $4K + O_2 \rightarrow 2K_2O$:	_____	_____
b) $2SO_2 + O_2 \rightarrow 2SO_3$:	_____	_____
c) $Mg + Br_2 \rightarrow MgBr_2$:	_____	_____
d) $NO_3^- + SO_3^{2-} \rightarrow NO_2^- + SO_4^{2-}$:	_____	_____

My Professor Says:

Do I get it? Let me check:

Oxidation Number

Oxidation number is the difference between normal valence electrons of the element and the number of electrons attributed to that element in a Lewis structure of the polyatomic ion or compound. We first give all the electrons in each bond to the more electronegative element in that bond. Then we subtract the number of electrons held by the element from its normal valence electrons (or the Group number). *Electronegativity* is the ability of an element to attract electrons to itself. Among chemically active elements, fluorine is the most electronegative, followed by those closest to it in the Periodic table. Consider the Lewis structure of HCN which is **H–C≡N**: *H* keeps $0e^-$, so its Ox # $= 1 - 0 = +1$; *C* keeps $2e^-$, so its Ox # $= 4 - 2 = +2$; *N* keeps $6e^- + 2e^-$, so its Ox # $= 5 - 8 = -3$

When given the Oxidizer/Reducer strength table (below) identify the oxidizer (or reducer) on the left and on the right. If the stronger oxidizer (or reducer) is on the left the reaction will proceed to a *LARGE extent*: If the stronger oxidizer (or reducer) is on the right the reaction will proceed to a *SMALL extent*. In many cases, reduction–oxidation reactions may involve H_2O as a reactant. Such reactions are not easily balanced by inspection because H_2O-related products may vary according to acidity or basicity of the medium. To balance these redox equations, we use the *half-reaction method* where we split the redox reaction into

Practice Questions

4.10. Use the table and predict if the reactions below will proceed to a LARGE or SMALL extent:

Oxidizer	Reducer
Ag^+	Ag
Fe^{3+}	Fe
Cu^+	Cu
Ni^{2+}	Ni
Cd^{2+}	Cd
Zn^{2+}	Zn

a) $Ni + Zn^{2+} \rightarrow Ni^{2+} + Zn$ _____

b) $3Cd + 2Fe^{3+} \rightarrow 3Cd^{2+} + 2Fe$ _____

c) $Cu^+ + Ag \rightarrow Cu + Ag^+$ _____

My Professor Says:

oxidation and reduction *half-reactions*. Steps in the half-reaction method for balancing an equation are:

1. Divide the skeleton reaction into two half-reactions, each containing the oxidized and reduced forms of one of the elements.

2. Balance the atoms and charges in each half-reaction as follows:

   a) First balance atoms other than O and H

   b) Balance O atoms by adding H_2O molecules

   c) Balance H atoms by adding H^+ ions

   d) Balance charge by adding electrons (e^-) to the left side in the reduction half-reaction AND to the right side in the oxidation half-reaction.

3. If necessary, multiply one or both half-reactions by an integer so that the number of electrons gained in the reduction half equals number of electrons lost in the oxidation half reaction.

4. Add the balanced half-reactions, and include physical states, that is, *s, l, g,* or *aq*

5. In acid solutions, add one H_2O to *both* sides of the equation acidic solutions. NOTE: Every H_2O and H^+ pair that appears on the *same* side of the equation form one hydronium ion, H_3O^+

Do I get it? Let me check:

Practice Questions

4.11. Balance the reduction-oxidation reaction below. Show all your work

$$ClO^- + I^- \rightarrow Cl^- + I_3^-$$

When it occurs in *basic* solution

My Professor Says:

6. In basic solutions, add one OH^- ion to *both* sides of the equation.
   NOTE: Every OH^- and H^+ ion pair that appears on the *same* side of the equation form one H_2O molecule.

Acid–Base Reactions

When given the Acid/Base strength table we can identify the acid on the left and on the right. If the stronger acid is on the left the reaction will proceed to a large extent: If the stronger acid is on the right the reaction will proceed to a small extent. The same is true if we compare the reactant and product bases.

Titration is a laboratory technique where two solutions react completely—one of known molarity and the other of unknown molarity. A balanced chemical equation of that reaction helps us to determine the unknown molarity. We gradually add a solution of known molarity to a fixed volume of the solution of unknown molarity. At the end point (when indicator color changes), we multiply volume by molarity to get moles added. We commonly use titration to determine molarity of unknown *acid* solution, for example, *15.00 mL of $H_3C_6H_5O_{7(aq)}$* using a *base* solution of known molarity, for example, *0.10 M $NaOH_{(aq)}$*, or *vice versa*. For these examples, the following overall reaction occurs:

$$H_3C_6H_5O_7(aq) + 3NaOH(aq) \rightarrow$$
$$Na_3C_6H_5O_7(aq) + 3H_2O(l)$$

Do I get it? Let me check:

Practice Questions

4.12. Use the strength table below to predict if the reactions that follow will proceed to a LARGE or SMALL extent:

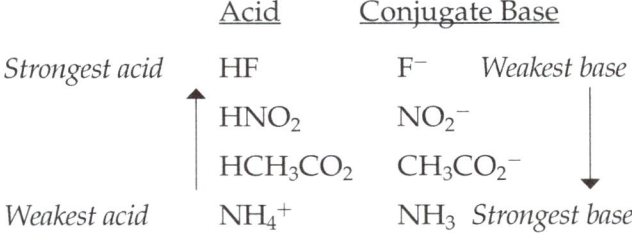

a) $HCH_3CO_2 + F^- \rightarrow$
   $HF + CH_3CO_2^-$ _____

b) $HCH_3CO_2 + NO_2^- \rightarrow$
   $HNO_2 + CH_3CO_2^-$ _____

c) $NH_4^+ + F^- \rightarrow$
   $NH_3 + HF$ _____

d) $NH_3 + HNO_2 \rightarrow$
   $NH_4^+ + NO_2^-$ _____

4.13. A 25.00 mL sample of KOH solution is titrated with 21.83 mL of 0.2120 M $HCl_{(aq)}$.

$KOH(aq) + HCl(aq) \rightarrow KCl(aq) + H_2O(l)$

What is the molarity of the KOH solution?

My Professor Says:

Do I get it? Let me check:

Suppose the volume of 0.10 mol/L NaOH required for complete reaction was 21.20 mL. How do we calculate molarity of 15.00 mL of $H_3C_6H_5O_{7(aq)}$ solution? Here is how:

$$\frac{\left[21.20\,mL \times \dfrac{1\,L\,Soln}{1000\,mL} \times \dfrac{0.10\,mol\,NaOH}{1\,L\,Soln} \times \dfrac{1\,mol\,H_3C_6H_5O_7}{3\,mol\,NaOH}\right]}{\left[15.00\,mL \times \dfrac{1\,L\,Soln}{1000\,mL}\right]}$$

$= 0.04711\,mol/L\,H_3C_6H_5O_7(aq)$

THESE ARE MY OWN NOTES
(from listening to lectures, watching YouTube videos, etc.)

Date: _____

Date: _____

Date: _____

Date: _____

I wish the professor could explain this: _____

Student's Practical Laboratory Experiment

ACID–BASE TITRATION

OBJECTIVE

The objective of this experiment is to determine the molarity of acetic acid ($HC_2H_3O_2$) and citric acid ($H_3C_6H_5O_7$) in unknown solutions by titration with standardized NaOH and phenolphthalein indicator. These are Brønsted–Lowry acids—they donate H^+ ion (or proton) to a base in solution.

The molarity of acetic acid, $HC_2H_3O_2$, in a sample of white vinegar will be determined in **Part A**:

$$HC_2H_3O_2(aq) + NaOH(aq) \rightarrow NaC_2H_3O_2(aq) + H_2O(l)$$

The molarity of citric acid, $H_3C_6H_5O_7$, in unknown solution will be determined in **part B:**

$$H_3C_6H_5O_7(aq) + 3\ NaOH(aq) \rightarrow Na_3C_6H_5O_7(aq) + 3\ H_2O(l)$$

LEARNING OUTCOMES

After this experiment, students will be able to:

1. Perform titration of acid with a base solution or *vice versa* as a lab technique
2. Determine the molarity of an acid in solution

SAFETY NOTES

Always wear safety goggles and shoes when doing experiments. <u>When finished, wash your hands with soap and water.</u>

PRE-LAB ASSIGNMENT

1. Write the following balanced chemical equations:

   a) *Molecular* equation between acetic acid, $HC_2H_3O_2$, and sodium hydroxide, NaOH:

   _____

   b) <u>Total ionic</u> equation between $HC_2H_3O_2$ (a weak acid) and NaOH:

   _____

   c) <u>Net ionic</u> between acetic acid, $HC_2H_3O_2$ (a weak acid), and NaOH:

   _____

   d) *Molecular* between citric acid, $H_3C_6H_5O_7$ (a weak acid), and sodium hydroxide, NaOH:

   _____

   e) <u>Net ionic</u> equation for the reaction between $H_3C_6H_5O_7$ (a weak acid) and NaOH:

   _____

2. Write the names of the glassware below:

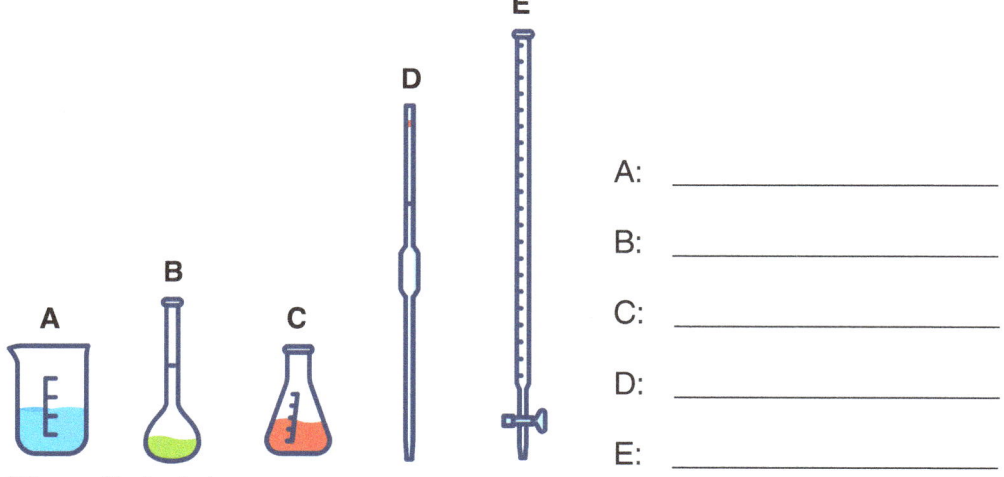

© Tsuyna/Shutterstock.com

A: _____

B: _____

C: _____

D: _____

E: _____

Materials Needed

1. Household vinegar, citric acid solution, 0.10 M NaOH and phenolphthalein indicator solution
2. A 5 mL buret, a 20 mL volumetric pipet, four 125 mL Erlenmeyer flask, 100 mL vol flask
3. A 250 mL Erlenmeyer flask with a cork, and a 50 mL beaker

Chemical Equations   **89**

EXPERIMENTAL PROCEDURE

Titration of Acetic acid with _____ mol/L NaOH solution

Part A: [Acid–Base Reaction: _____]

1. Pipet 20.00 mL vinegar into 100.00 mL vol. flask. Fill up to the mark with DI water. Mix well.

2. Pipet 20.00 mL dilute solution (1) into the first 125 mL Erlenmeyer flask. Add 3 drops phenolphthalein. Add NaOH from the buret while swirling until a permanent pink. Record end-point volume. Repeat for Runs 1–3.

	Test Run	Run 1	Run 2	Run 3
Final buret reading				
Initial buret reading				
Vol. (**mL**) NaOH delivered				
Use Dimensional Analysis to Convert units:				
Vol. NaOH delivered (**L**)				
mol NaOH delivered				
*Calculation Set-up to Convert **mL** NaOH to **mol** $HC_2H_3O_2$:*				
mol $HC_2H_3O_2$				
Vol. $HC_2H_3O_2$ solution (**mL**)				
Vol. $HC_2H_3O_2$ solution (**L**)				
*Calculation Set-up to Convert **mL** NaOH to **mol/L** $HC_2H_3O_2$:*				
Molarity (**mol/L**) $HC_2H_3O_2$				
Avg Molarity $HC_2H_3O_2$ (**mol/L**) =				
Use the dilution equation ($M_1V_1 = M_2V_2$) to calculate the *molarity* of Vinegar				

Titration of Citric Acid with _____ **mol/L** NaOH solution

Part B: [Acid–Base Reaction: _____]

Pipet 20.00 mL citric acid solution into the first 125 mL Erlenmeyer flask. Add 3 drops phenolphthalein. Add NaOH from the buret while swirling until a permanent pink. Record end-point volume. Repeat for Runs 1–3.

	Test Run	Run 1	Run 2	Run 3
Final buret reading				
Initial buret reading				
Vol. (**mL**) NaOH delivered				
Use Dimensional Analysis to Convert units:				
Vol. NaOH delivered (**L**)				
mol NaOH delivered				
*Calculation Set-up to Convert **mL** NaOH to **mol** $HC_2H_3O_2$:*				
mol $H_3C_6H_5O_7$				
Vol. $H_3C_6H_5O_7$ solution (**mL**)				
Vol. $H_3C_6H_5O_7$ solution (**L**)				
*Calculation Set-up to Convert **mL** NaOH to **mol/L** $H_3C_6H_5O_7$:*				
Molarity (**mol/L**) $H_3C_6H_5O_7$				
Avg Molarity $H_3C_6H_5O_7$ (**mol/L**) =				

—oOo—   **END OF LAB REPORT**

THESE ARE MY OWN NOTES
(from listening to lectures, watching YouTube videos, etc.)

Date: _____

Date: _____

Date: _____

Date: _____

I wish the professor could explain this: _____

Chemical Equations

THE ATOM AND ITS INTERNAL STRUCTURE — 5

Development of the Atomic Theory

Back, many many years ago, in 450 B.C., a Greek-speaking man named Democritus proposed that if you divided a piece of a pure sample (of a given element) into two parts, and then divide the one half, and half of that half, and so on, you would end up with a piece that cannot be divided further. Democritus called that indivisible piece, an "atom."

Viewed this way, it only made sense at that time to imagine an atom as a simple particle. However, some centuries later, Joseph Thomspon discovered that atoms have *electrons*. Soon after, Ernest Rutherford discovered that an atom has a dense core which he called a *nucleus*, and proposed that the nucleus was positively charged. Therefore electrons, which are never stationary but move continually, were held in the immediate vicinity of the nucleus—around it. Niels Bohr proposed that for a hydrogen atom, an electron orbits the nucleus. However, Bohr's model did not readily apply to atoms with many electrons. Experiments prove that any charge that precesses around a center of opposite charge always spirals in and collapses. For these reasons, Niels Bohr's model of the atom needed to be refined. Other scientists also investigated the physical arrangement of electrons in an atom.

Max Plank discovered that when an atom absorbs energy, it absorbs it in fixed quantities. One such quantity is referred to as a "quantum." More such quantities are "quanta."

Atomic Models

Solid sphere model (Dalton, 1803) • Plum pudding model (Thomson, 1897) • Nuclear model (Rutherford, 1911) • Planetary model (Bohr 1913) • Quantum model (Schrödinger, 1926)

© N.VinothNarasingam/Shutterstock.com

My Professor Says:

Do I get it? Let me check:

Atomic Orbitals and Energy Levels

To express Plank's observation, some authors say energy is "quantized." The energy absorbed makes electrons, although still confined to the immediate vicinity of the nucleus, to be held at greater but specific (or discrete) distances from the nucleus. In contrast to a planar orbit as proposed by Bohr, a scientist named Erwin Schrodinger proposed that electrons oscillate within specific differentiated spatial volumes called orbitals which are centered on the nucleus as shown on the right. Due to its high speed, when an electron oscillates, it appears as a standing wave that outlines its oscillation boundaries, that is, its orbital. Schrodinger predicts spatial oscillation volumes (orbitals) with unique shapes and sizes, when placed on *xyz*–axes. The orbitals can be represented as shown below. Each orbital can be occupied by one, or a maximum of 2 electrons.

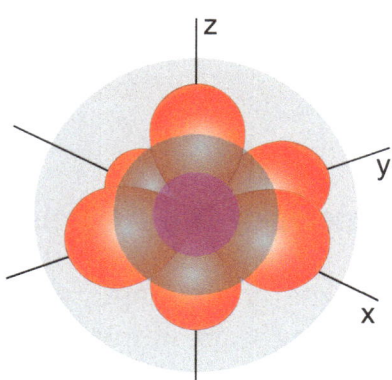

Some standing waves or orbitals are spherical. Others are shaped like dumbbell-weights. In this diagram, the number of spherical orbitals is _____ including the outermost gray one. The number of dumbbell-shaped orbitals is _____.

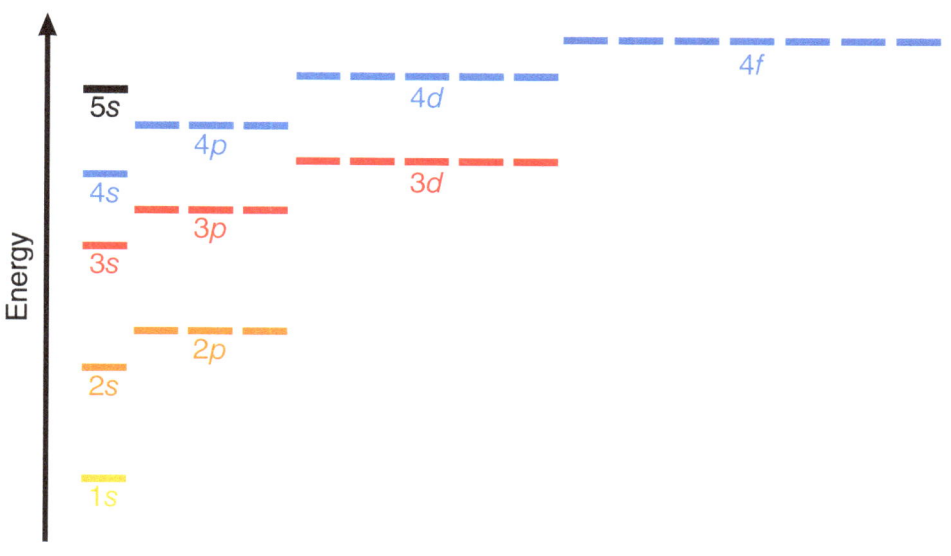

My Professor Says:

Quantum Numbers

To identify a specific orbital, we first *identify its energy level* designated as *principal quantum number "n."* Then we *identify its sublevel* designated as *secondary quantum number "l"* within the energy level where it is found. Next, we *identify its orientation on the xyz axes* designated as *magnetic quantum number "m_l."* In addition to the 4s-, 4p-, and 4d-orbitals, the fourth energy level, n = 4, has seven 4f-orbitals. The shapes of the various orbitals outlined by electron oscillation around the nucleus of an atom are shown below. The lowest energy level, n = 1, has only one spherical 1s- orbital; its sublevel is $l = 0$, and its orientation is $m_l = 0$. The second energy level, n = 2, has one spherical 2s-orbital and three dumbbell shaped 2p-orbitals—one along the x-axis (p_x), the second along the y-axis (p_y), and the third along the

Do I get it? Let me check:

Practice Questions

5.1. The energy level $n = 4$ consists of __one(1)__ s-orbital, _____ p-orbitals, _____ d-orbitals, and _____ f-orbitals."

5.2. Circle orbital(s) represented by this picture

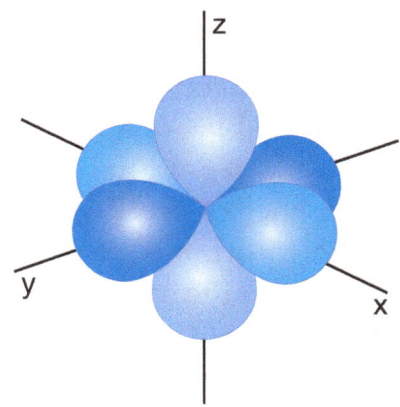

a) $d_{x^2-y^2}$ d) d_{z^2}
b) d_{xz} and d_{yz} e) f-orbital
c) p_z, p_x, and p_y

l	0	1			2					3						
m_l	0	-1	0	1	-2	-1	0	1	2	-3	-2	-1	0	1	2	3
n	s	p_x	p_y	p_z	d_{xy}	d_{xz}	d_{z^2}	d_{yz}	$d_{x^2-y^2}$	$f_{x(x^2-3y^2)}$	f_{xz^2}	f_{xz^2}	f_{z^2}	f_{yz^2}	f_{z^2}	$f_{y(3x^2-y^2)}$
1	●															
2	●	●	●	●												
3	●	●	●	●	●	●	●	●	●							
4	●	●	●	●	●	●	●	●	●	●	●	●	●	●	●	●
5	●	●	●	●	●	●	●	●	●							
6	●	●	●	●												
7	●															

My Professor Says:

z-axis (p_z). The third energy level, $n = 3$, has one spherical 3s-orbital, three dumbbell shaped 3p-orbitals—p_x, p_y, p_z, and five 3d-orbitals. Of the five 3d-orbitals, one extends between x and y axes in the xy-plane (d_{xy}); the second between x and z axes in the xz-plane (d_{xz}); the third between y and z axes in the yz-plane (d_{yz}); the fourth called the $d_{(x^2-y^2)}$-orbital, along the x and the y axes; and the fifth called d_{z^2}-orbital, along the z-axis and equatorially around the z-axis. The s-orbital is always designated $l = 0$, a p-orbital $l = 1$, a d-orbital $l = 2$, an f-orbital $l = 3$, and so on. For each energy level, we designate the m_l values to orbitals of a given sublevel starting from the center orbital, as $m_l = 0(central)$; $m_l = -1(first\ left)$, $m_l = +1(first\ right)$, $m_l = -2(second\ left)$, $m_l = +2(second\ right)$, and so on. If one or two electrons occupy an orbital, we designate the one with a spin quantum number, $m_s = -½$ and the other $m_s = +½$. In general, therefore, four quantum numbers, n, l, m_l, and m_s are used to describe a specific orbital and/or electron in an atom.

Electron Configuration

Due to electro-electron repulsions which lead to greater radial separations between orbitals at high energy levels ($n \geq 3$), some electrons in the next higher energy level tend to penetrate into the lower energy levels. As a result, there are overlaps between orbitals at higher energy levels ($n \geq 3$). In order to keep track of these overlaps, we list the orbitals (as sublevels)

Do I get it? Let me check:

Practice Questions

5.3. Considering their size and xyz orientation, match each atomic orbital (A, B, or C) below with the quantum numbers that follow:

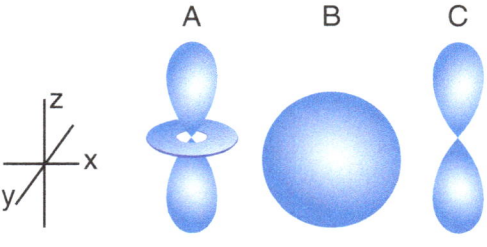

a) $n = 3$ and $l = 2$ _____

b) $n = 2$ and $l = 0$ _____

c) $n = 3$ and $l = 1$ _____

5.3. What is a possible set of quantum numbers for the <u>unpaired electron</u> in the orbital box diagram below?

[Ar] ↑↓|↑↓|↑↓|↑↓|↑↓ ↑↓ ↑↓|↑↓|↑↓
        3d   4s   4p

a) $n = 1, \ell = 1, m\ell = -1, m_s = +½$

b) $n = 3, \ell = 2, m\ell = -1, m_s = -½$

c) $n = 4, \ell = 2, m\ell = -2, m_s = +½$

d) $n = 4, \ell = 0, m\ell = 0, m_s = +½$

e) $n = 4, \ell = 1, m\ell = -1, m_s = +½$

5.4. The maximum number of electrons that can occupy the $n = 4$ energy level is _____

My Professor Says:

according to the *"Aufbau"* diagram below (insert) when we complete the electron configuration of an element. For example, the electron configuration for nickel, Ni, is $1s^2 2s^2 2p^6 3s^2 3p^6 4s^2 3d^8$ where 4s has lower energy than 3d and is therefore filled before 3d:

↑↓   ↑↓ ↑↓ ↑↓ ↑ ↑
4s   3d

↑↓   ↑↓ ↑↓ ↑↓
3s   3p

↑↓   ↑↓ ↑↓ ↑↓
2s   2p

↑↓
1s

Electron configurations with completely filled and half-filled *d*-subshell are relatively stable. So, instead of being one electron shy of either completely filling or half-filling the *d*-subshell, one electron is moved over from the 4s orbital. For example, the electron configuration for silver, Ag, is $1s^2 2s^2 2p^6 3s^2 3p^6 4s^2 3d^{10} 4p^6 5s^1 4d^{10}$ where 4d is completely filled and 5s is half-filled.

Do I get it? Let me check:

Practice Questions

5.5. The following orbital diagram corresponds to which element?
_____

↑↓   ↑↓ ↑↓ ↑
3s   3p

↑↓   ↑↓ ↑↓ ↑↓
2s   2p

↑↓
1s

5.6. Write the full *spdf*-electron configuration of manganese: _____
_____

5.7. Write the <u>full</u> electron configuration of plutonium: _____
_____

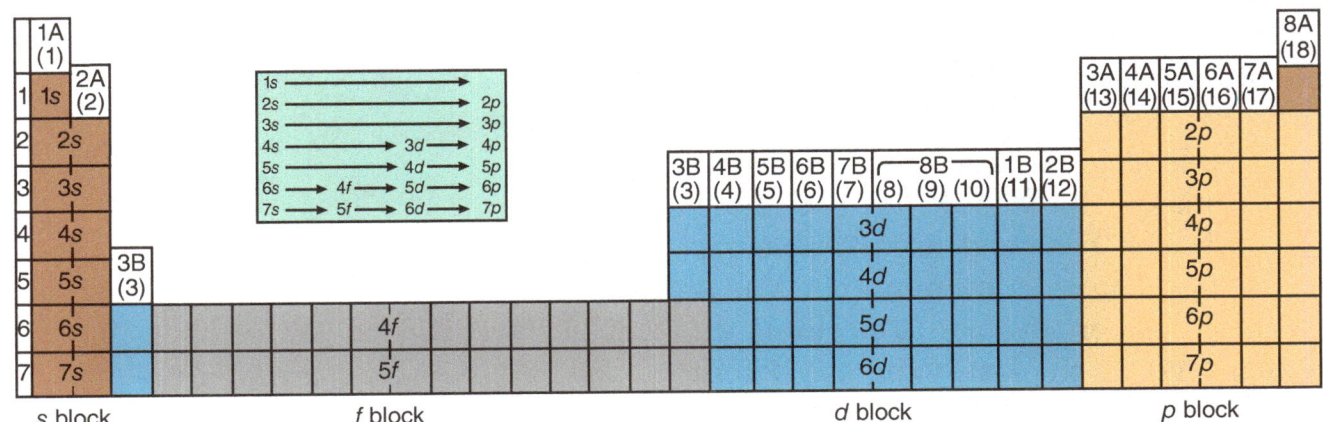

My Professor Says:

Filling orbitals of low energy level before those of higher energy levels as we have just discussed is the first rule called the *Aufbau principle*. As shown in the diagram above, the Aufbau diagram shadows the periodic table. Ultimately, you should use the periodic table as an aid for building electron configurations. The second rule called *Hund's rule* requires us to fill in a single electron in each orbital of the same energy first and then fill in the second pairing electron. The third rule called *Pauli's exclusion principle* requires us to show that the second electron in each orbital has opposite spin. Once we have built the energy level diagram following Aufbau we can fill in all the electrons a given atom has following Hund's rule and exclusion principle. In fact, we can identify the specific element if we have its electron-filled energy level diagram, also referred to as the *electron configuration*.

We do not always draw an energy level diagram in order to show the electron configuration of an element. Instead, we may write it down using full "*spdf*" notation or a condensed "*noble-gas core*" notation. For example, electron configuration for aluminum atom (Al, element 13 on the Periodic table) may be written as $1s^2 2s^2 2p^6 3s^2 3p^1$ which is a full "spdf" notation, where the first number represents the energy level, the letter represents the orbital, and the exponent represents the number of electrons filling the orbital(s). The condensed "noble-gas core" notation for aluminum (Al) is $[Ne]3s^2 3p^1$ where the first 10 electrons are represented by the last noble gas before the element, in this case neon, Ne.

Do I get it? Let me check:

Practice Questions

5.8. An element has this "noble-gas core" electron configuration:

$[Ar]$ ↑ | ↑ ↑ ↑ ↑ ↑
    4s   3d

What is the name of the element?

_____

5.9. Which element has the abbreviated electron configuration $[Ar]4s^2 3d^8$?

_____

My Professor Says:

Simple ions are atoms which lost or gained electrons. In order to show the correct electron configuration for a given ion, we always complete the electron configuration for an atom of the given element first and then, in the case of cations, remove the electron lost from the outermost orbital(s) in the highest energy level. In the case of anions we simply add the electrons gained.

Periodic Trends

Atomic size increases down a group as the principal quantum number, *n*, increases, and decreases left to right across a period as effective nuclear charge (Z_{eff}) increases; Z_{eff} = total electrons − core electrons. *Ionization energy* (IE) trend is exactly opposite atomic size. Although IE can be viewed as complete removal of one electron from the atom, it is experimentally determined as energy to remove one mole of electrons from one mole of atoms. Electron affinity (EA), the opposite of IE, has many exceptions. For covalently bonded atoms, electronegativity, that is, electron-holding strength increases up a group and toward fluorine (F).

Cations are smaller than their parent atoms while anions are larger. If isoelectronic, that is, having equal number of electrons, *ionic radius* decreases from high negative to a high positive charge (−3 > −2 > −1 > +1 > +2 > +3).

Do I get it? Let me check:

Practice Questions

5.10. Fill in the electrons (↑ or ↓) in the box diagram below to show the electron configuration of cobalt (III) ion.

[Ar] ☐ ☐☐☐☐☐
     4s   3d

5.11. Arrange the following species in order of *decreasing* size. Explain:

a) Ca^{2+}, Sr^{2+}, Mg^{2+}

_____

_____

_____

b) K^+, S^{2-}, Cl^-

_____

_____

_____

c) Au^+, Au, Au^{3+}

_____

_____

_____

THESE ARE MY OWN NOTES
(from listening to lectures, watching YouTube videos, etc.)

Date: _____

Date: _____

Date: _____

Date: _____

I wish the professor could explain this: _____

Student's Practical Laboratory Experiment

REDUCTION–OXIDATION (REDOX) REACTIONS

OBJECTIVE

To demonstrate reduction–oxidation reactions where one reactant loses some of its electrons, and the same electrons are then gained by the other reactant.

Oxidation is any process which increases the oxidation state of an element, or makes its charge more positive, due to the loss of electrons.

Reduction is any process which reduces the oxidation state of an element, or makes its charge more negative, due to the gaining of electrons.

In these reactions, one reactant is the *oxidizing agent (or oxidizer)* and the other reactant is the *reducing agent (or reducer).*

- The oxidizing agent oxidizes something else by accepting electrons from it, and it is reduced.
- The reducing agent reduces something else by giving it electrons, and it is oxidized.

In this experiment, you will perform simple tests to determine the relative reactivity of several metals (as reducing agents). You will rank the reactivity of these metals by comparing their reaction with water first and then with an acid solution.

LEARNING OUTCOMES

After this experiment, students will be able to:

1. Identify and rank chemicals according to their oxidizing or reducing strength
2. Predict whether a reduction–oxidation reaction will occur as written or not.

SAFETY NOTES

Group 1A metals react spontaneously with air or water, producing fire and large amounts of gas. Keep these stored in mineral oil, and use small amounts. Always wear safety goggles and shoes when doing experiments. <u>When finished, wash your hands with soap and water.</u>

PRE-LAB ASSIGNMENT

1. Your instructor will demonstrate the reaction of potassium (**K**) metal with water (**H₂O**). This reaction generates a large amount of energy as well as hydrogen gas. Balance this equation and answer the questions below.:

   ____ K(s) + ____ H$_2$O(l) → ____ K$^+$(aq) + ____ OH$^-$(aq) + ____ H$_2$(g)

		Oxidation Numbers			Reduction/Oxidation?
a) Potassium:	K	_____	→	K$^+$ _____	_____
b) Hydrogen:	H$_2$O	_____	→	H$_2$ _____	_____

2. Consider the following reaction: **2 Al(s) + 3 Cl$_2$(g) → 2 Al^{3+}(aq) + 6 Cl$^-$ (aq)**

   a) Which reactant is oxidized? _____

   b) Which reactant is reduced? _____

   c) Which reactant is the reducing agent? _____

   d) Which reactant is the oxidizing agent? _____

   e) If the reaction occurs as written, which is the strongest reducing agent in the equation? _____

   f) How many electrons were transferred in this process? _____

Materials Needed

1. Some potassium, magnesium, calcium, copper, zinc metals
2. Distilled H$_2$O and 1 M hydrochloric acid, HCl(aq) solution
3. About eight (8) 20 mL test tubes

The Atom and Its Internal Structure **107**

EXPERIMENTAL PROCEDURE

Beginning with the reaction of potassium and water demonstrated by your instructor, write what you observe which indicates that a reaction is taking place.

Add 2–3 mL of deionized water to each tube. Record your observations, especially if a gas is being generated, and the rate at which the gas is being generated relative to the other metals. You can test each test tube with a strip of pH paper to see if the reaction generated hydroxide ions and record the result. You may have to wait more than a minute to observe a reaction as some reactions occur very slowly.

If no reaction is observed, write "No Reaction." If no reaction happens, the metal and water were too weak of a redox pair to produce their stronger conjugates.

1. **Reactions of metals with pure H_2O**

Metal	Observations	Write the balanced equation
K		
Ca		
Cu		
Mg		
Zn		

2. **Reactions of metals with 1 M HCl**

    In the table below write the chemical symbols, in alphabetical order, for the elements which you did not observe reacting in the previous section. Pour out the water from these test tubes and add 2–3 mL of 1 M HCl solution instead of pure water and observe any reaction.

Caution: *To avoid excessive heat and gas formation, add 1M HCl only to metals that did NOT react with water.* Recognize that this aqueous solution of HCl contains only hydronium ions and chloride ions, as it is a strong acid that has reacted completely with water.

Metal	Observations	If reaction occurs, write the equation

3. **Build a Redox Table:**

   Based on your observation of how strong each reducing agent was, create a strength table of all of the reducing agents (including the conjugate reducing agents) and the oxidizing agents (including any conjugate oxidizing agents).

   Note: For the two oxidizing agents you used, each has a pair of conjugate reducing agents.

   The first reaction was $2\,H_2O \rightarrow H_2 + 2\,OH^-$, and the second reaction was $2\,H_3O^+ \rightarrow H_2 + 2\,H_2O$. In each of these equations, the reactant is the oxidizing agent, while the products are a pair of reducing agents that go together in the *strength table*.

Reducing agents	**Oxidizing agents**
[Strongest(top) to Weakest]	*[Weakest(top) to strongest]*
_____	_____
_____	_____
_____	_____
_____	_____
_____	_____
_____	_____
_____	_____

4. **Use the *strength table* to predict favored reactions**

   Now that you have built your strength table, use it to determine which of the following reactions will go to completion, and which one will not happen at all.

   <u>*Circle the correct choice*</u>

   1. $Cu(s) + Zn^{2+}(aq) \rightarrow Zn(s) + Cu^{2+}(aq)$ **A)** full extent   OR   **B)** no reaction
   2. $Zn(s) + Cu^{2+}(aq) \rightarrow Cu(s) + Zn^{2+}(aq)$ **A)** full extent   OR   **B)** no reaction

5. **Test your predictions:**

   Put a piece of copper metal in an aqueous solution of zinc ions [1.0 M $Zn(NO_3)_2$] and a piece of zinc metal in an aqueous solution of copper ions [1.0 M $Cu(NO_3)_2$].

   Which reaction proceeds as written? Explain _____

   _____

   _____

   _____

—o0o—   **END OF LAB REPORT**

THESE ARE MY OWN NOTES
(from listening to lectures, watching YouTube videos, etc.)

Date: _____

Date: _____

Date: _____

Date: _____

I wish the professor could explain this: _____

MOLECULAR STRUCTURES 6

Lewis Symbols and Chemical Bonds

An atom of a Group 1A element (in the Periodic Table) has *one* valence electron. An atom of a Group 2A element has *two* valence electrons; that of a Group 3A has three valence electrons, that of a Group 4A element, *four*, and so on, through Group 8A. Going across a Period, from one group to the next, atoms have increasing number of valence electrons as illustrated by atoms of Period 2 elements shown below.

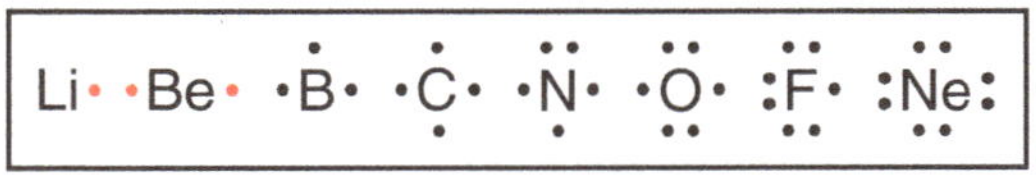

Li· ·Be· ·B· ·C· ·N· ·O· :F· :Ne:

A Lewis symbol, therefore, is simply the element symbol surrounded by as many valence electrons as its group number in the Periodic Table. Note that valence electrons (up to the fourth one) are placed separately in the West, East, South, North (W.E.S.N) positions around the element symbol and pairing starts with the fifth, through the eighth electron. We fill electrons in this order to comply with the *Octet Rule*. According to this rule, atoms are most stable when they have eight (8) valence electrons. The chemical stability of noble gases is attributed to their having octet, that is, 8 valence electrons. In order to attain

Practice Questions

6.1. Formation of *covalent bonds*

	Lewis Symbol	Draw Molecule Formed with H-atom
Carbon		
Sulfur		
Phosphorus		

117

My Professor Says:

similar stability, metal atoms shed their valence electrons and become cations with 8 electrons in their outermost shell while nonmetal atoms attract as many electrons as they need to make 8 valence electrons, and by doing so become anions.

In most cases, a nonmetal atom's need for electrons is met when it couples its unpaired electron(s) with those of other nonmetal atoms thus forming a covalent bond. This happens when the two atoms come close enough to each other that their respective regions of nuclear pull overlap. Then, each atom's unpaired electron, while held tight by the nucleus, becomes attracted to the nucleus of the adjacent atom. So, a single covalent bond consists of two electrons shared between two atoms. It is also called a bonding pair, and it is commonly represented by a single line between the element symbols. Other electron pairs which do not form bonds are called lone pairs. If a nonmetal atom attracts electrons from elsewhere and becomes negatively charged, it can be electrostatically attracted to cations, forming an ionic bond. Lewis representation of a compound with ionic bonds (i.e., an ionic compound) is simply the Lewis Symbols in square brackets and with charge as superscript of the cation and the anion adjacent to each other, without any line joining the ions. Note that ionic compounds are held together by electrostatic attractions rather than a covalent bond.

Do I get it? Let me check:

Practice Questions

6.2. Formation of *ionic bonds*

	Lewis Symbol	Draw Formula Unit Formed with F atom
Lithium		
Magnesium		
Aluminum		

Building Lewis Structures

A Lewis representation consists of a central atom and bonded atoms. While some molecules and polyatomic ions have only single bonds in their Lewis representations, others have double and triple bonds. How do these double and triple bonds come about? Others yet, have two or more central atoms connected in a linear way, or branched. Correct Lewis representations of any chemical formula, for example, H_2O, or CNO^-, or NH_4^+, or CH_3CONH_2 are arrived at by following specific rules. The rules can be outlined as follows:

1. Calculate the total number of valence electrons. To do this we simply add the group numbers of each atom in the structure, and *for negatively charged ions add charge* BUT <u>for positively charged ions subtract charge.</u>

    - Since H_2O has *two* Group 1 atoms, *Hydrogens*, and *one* Group 6 atom, *oxygen*, and *no charge*, the **total number of valence electrons in H_2O = 2(1e) + 1(6e) + 0 = 8e.**

    - For the cyanate ion, CNO^-, the negative charge indicates that there is one additional electron. So, the **total number of electrons in CNO^- = 1(4e) + 1(5e) + 1(6e) + (1e) = 16e.**

    - For the ammonium ion, NH_4^+, the positive charge indicates that the number of electrons is less by the same number as the charge (i.e., by 1). So, the **total number of electrons in NH_4^+ = 1(5e) + 4(1e) − 1e = 8e.**

My Professor Says:

2. Select the central atom(s) using the recommendations below in order. Many chemical formulas have one central atom except *those with a subscript on the central atom*, for example, C_4H_{10} or those showing functional groups such as CH_3CONH_2 where $-CH_3$, $-\overset{O}{C}-$, and $-NH_2$ are distinct groups, each with its own central atom. Central atoms of these groups are always joined to each other. For Lewis representations of any given formula you may choose:

   - the least electronegative atom. This is usually furthest away from fluorine in the Periodic Table or, if at equal radial distance from fluorine, it is the one in the bottom period.

   - the atom (element) with largest number of unpaired electrons in its Lewis symbol, that is, one which forms most bonds.

   - the first element in the chemical formula, except when it is hydrogen which never serves as the central element.

3. Distribute total valence electrons as single bonds (bonding pairs) first and then as lone (nonbonding) pairs around every surrounding atom until each has noble gas configuration.

4. If there are more electrons after giving each surrounding atom has noble gas configuration, place the remaining electrons as lone pairs on the central atom.

Do I get it? Let me check:

Practice Questions

6.3. Formation of *Noble gas Configuration*

Formula	Tot. Valence Electrons	Lewis Structure

NH_3	1(5e) + 3(1e) = 8e	H—N̈—H
		    |
		    H

H_2S		
CO_2		
CO_3^{2-}		
$CH_3(CO)OH$		

My Professor Says:

5. If total electrons run out before every atom has noble gas configuration, convert one lone pair into a bonding pair; continue as required until noble gas configuration is attained.

6. Put any polyatomic ion in brackets and write the charge outside as a superscript. Draw all resonance structures if possible. Draw all isomers if possible.

7. Calculate formal charge for each element in each structure as follows:

   - First, split each bond equally so each atom gets half of the electrons
   - Formal charge = (Group No.) − (No. of electrons around the atom)

Do I get it? Let me check:

Practice Questions

6.4. Lewis Structure with *minimum Formal Charges*

Formula	Tot. Valence Electrons	Lewis Structure
PCl_5		
SF_4		
SF_6		
ICl_4^-		
$XeOF_4$		
XeF_4		

My Professor Says:

Resonance, Formal Charge, and Isomers

It is possible to generate more than one Lewis representation of a given formula. The central atom may have a double (or a triple) bond with one atom in one representation, and with the other atom in another representation. Such Lewis representations which differ only in arrangement of electrons are called *resonance structures*. Please note that resonance structures are different representations of one molecule or polyatomic ion.

For example, the azide ion, N_3^-, has three *reasonable* resonance structures which must be shown with double-headed arrows called *resonance arrows*, as follows:

$$[:\ddot{N}=N=\ddot{N}:]^- \leftrightarrow [:\ddot{N}-N\equiv N:]^-$$
$$(1) \qquad \qquad (2)$$
$$\updownarrow$$
$$[:N\equiv N-\ddot{N}:]^-$$
$$(3)$$

Reasonable resonance structures exclude any structure where elements are shown to form *atypical* bonds in addition to those that are typical. For example, a structure with triple-bonded oxygen is atypical if there are others where oxygen is single- and/or double bonded. Note that the true structure of a molecule or a polyatomic ion being modeled is believed to be a hybrid of the individual resonance structures. In order to identify one preferred resonance structure that is regarded as important, atoms are assigned a hypothetical property called formal charge (*FC*) which is given by:

Do I get it? Let me check:

Practice Questions

6.5. Formation of *Resonance Structures* [Formal charges in structures **1**, **2**, and **3** of the azide ion, N_3^-]:

No.	Formal Charge (FC)			Which structure is most important? Explain
	Left N	Middle N	Right N	
1	$5-6=$ -1	$5-4=$ $+1$	$5-6$ $=-1$	
2	$5-7=$ -2	$5-4=$ $+1$	$5-5$ $=0$	
3	$5-5$ $=0$	$5-4=$ $+1$	$5-7$ $=-2$	

My Professor Says:

FC_{atom} = (Group No.) − (No. of electrons assigned the atom)

Bonding electrons are always divided and each of the two atoms bonded to each other is assigned half of bonding electrons. The sum of *FC*s of all atoms in a given Lewis representation is zero for molecules or equal to the charge for polyatomic ions. The resonance structure chosen is one in which (**1**) negative formal charge is on the most electronegative atom, and (**2**) individual atoms have formal charges of zero or closest to zero.

Formulas with two or more central atoms, not written to show functional groups, usually represent different compounds (isomers). For example, $CH_3(CH_2)_2OH$ represents 1-propanol (with the OH group at a terminal carbon) ONLY whereas C_3H_8O represents 1-*propanol*, and *isopropanol* (with the OH group on the second carbon) as well as *ethyl-methyl ether* with the O-atom between the ethyl (CH_3CH_2) and methyl (CH_3) groups. In this case, we have a *different arrangement of the central atoms*. Such Lewis structures are called <u>structural isomers</u>, and each represents a different compound. Other types of isomers are *geometric* (or *cis-* and *trans*) isomers, where identical functional groups are fixed either on the same side (*cis-*) or opposite sides (*trans-*) of a double bond between two adjoining central atoms. *Chiral* (or *enantiomeric*) isomers are those in which a given *carbon* central atom has four different functional groups bonded to it.

Do I get it? Let me check:

Practice Questions

6.6. Draw *Reasonable Resonance Structures* for each Formula and show formal charges.

Formula	All Resonance Structures POSSIBLE
N_2O	
$HOCO_2^-$	
SCN^-	

My Professor Says:

Three-Dimensional (3D) Shapes and Drawings

Groups of electrons, whether arranged as a single, double, or triple bond, or as a lone pair, repel each other such that they extend away from the central atom, at specific angles relative to each other in the three-dimensional (3D) space around the central atom. If the structure has only TWO electron groups, the farthest they can be from each other is on opposite sides of the central atom, that is, 180° apart, making a line or a *Linear* arrangement. The farthest THREE electron groups that are attached to a central atom can be from each other is 120° apart, extending to the corners of a triangle thus making a *Trigonal Planar* arrangement. FOUR electron groups will be 109.5° apart, making a *Tetrahedral* arrangement. FIVE electron groups make two pyramids, one above and the other inverted below a triangular plane—a *Trigonal bipyramidal* arrangement. One of the five groups, extends up from the plane at 90° angle and another group extends down (in the opposite direction from the plane) at 90° angle, while three groups making a triangular plane are 120° apart. Note that lone pairs repel the most and therefore, prefer to be at 120° away from each other. [Evidently, the name *Trigonal bipyramidal* was not coined by the Greek. Were that the case, we would be talking of a "hexahedral" arrangement, since it follows tetrahedral.] SIX electron group make an *octahedral* arrangement where two extend

Do I get it? Let me check:

Practice Questions

6.7. Draw **all** possible *structural isomers* for the formula, $C_2H_4O_2$

Skeletons of central atoms ONLY	POSSIBLE Lewis Structures

My Professor Says:

up at 90° from a square plane made by other four groups while the other group extends down also at 90° from the plane. Groups making the square plane are also at 90° angles from each other.

This repulsion of electron groups is described by Valence Shell Electron **Group** Repulsion (VSEGR) theory, which predicts five basic arrangements, regarded as *parental arrangements (or shapes)* with "ideal" bond angles for the respective geometric arrangements specified above. Why are these regarded as parental shapes? The actual shapes of molecules are observed through a laboratory technique called x-ray crystallography which can visualize positions of atoms only and not the electrons. Thus, bonding pairs can be assumed to be located between atoms while a lone pair is seen simply as a space formed when electron groups repel each other. Therefore, in naming molecular shapes, we consider only the shape outlined by atoms. For example, if one of the three groups in a trigonal planar arrangement is a lone pair, the molecular shape is <u>Angular</u> or <u>Bent</u> (with less than 120° angle). If a tetrahedral arrangement has two lone pairs, the molecule will also be <u>Angular</u> or <u>Bent</u> (with less than 109.5° angle). A tetrahedral arrangement with one lone pair gives a <u>Trigonal Pyramidal</u> molecular shape (with less than 109.5° angle). In a trigonal bipyramidal electron arrangement, lone pairs lie preferably in the planar positions which have relatively more room; therefore, two lone pairs make a

Do I get it? Let me check:

Practice Questions

6.8. If **A** = central atom, **X** = bonded atom, **E** = lone pair, what will be the molecular geometry (or shape outlined by nuclei of bonded atoms)?

VSEGR notation	No. of e-Groups	Motif/Parental (e-arrangement)	Molecular Geometry (made by nuclei)	Example
AX₂	2	Linear		CO_2
AX₃	3	Trigonal Planar		SO_3
AX₂E	3	Trigonal Planar		SO_2
AX₄	4	Tetrahedral		CH_4
AX₃E	4	Tetrahedral		NH_3
AX₂E₂	4	Tetrahedral		H_2O
AX₅	5	Trigonal bipyramidal		PCl_5
AX₄E	5	Trigonal bipyramidal		SF_4
AX₃E₂	5	Trigonal bipyramidal		ClF_3
AX₂E₃	5	Trigonal bipyramidal		ICl_2^-
AX₆	6	Octahedral		SF_6
AX₅E	6	Octahedral		BrF_5
AX₄E₂	6	Octahedral		XeF_4

My Professor Says:

Do I get it? Let me check:

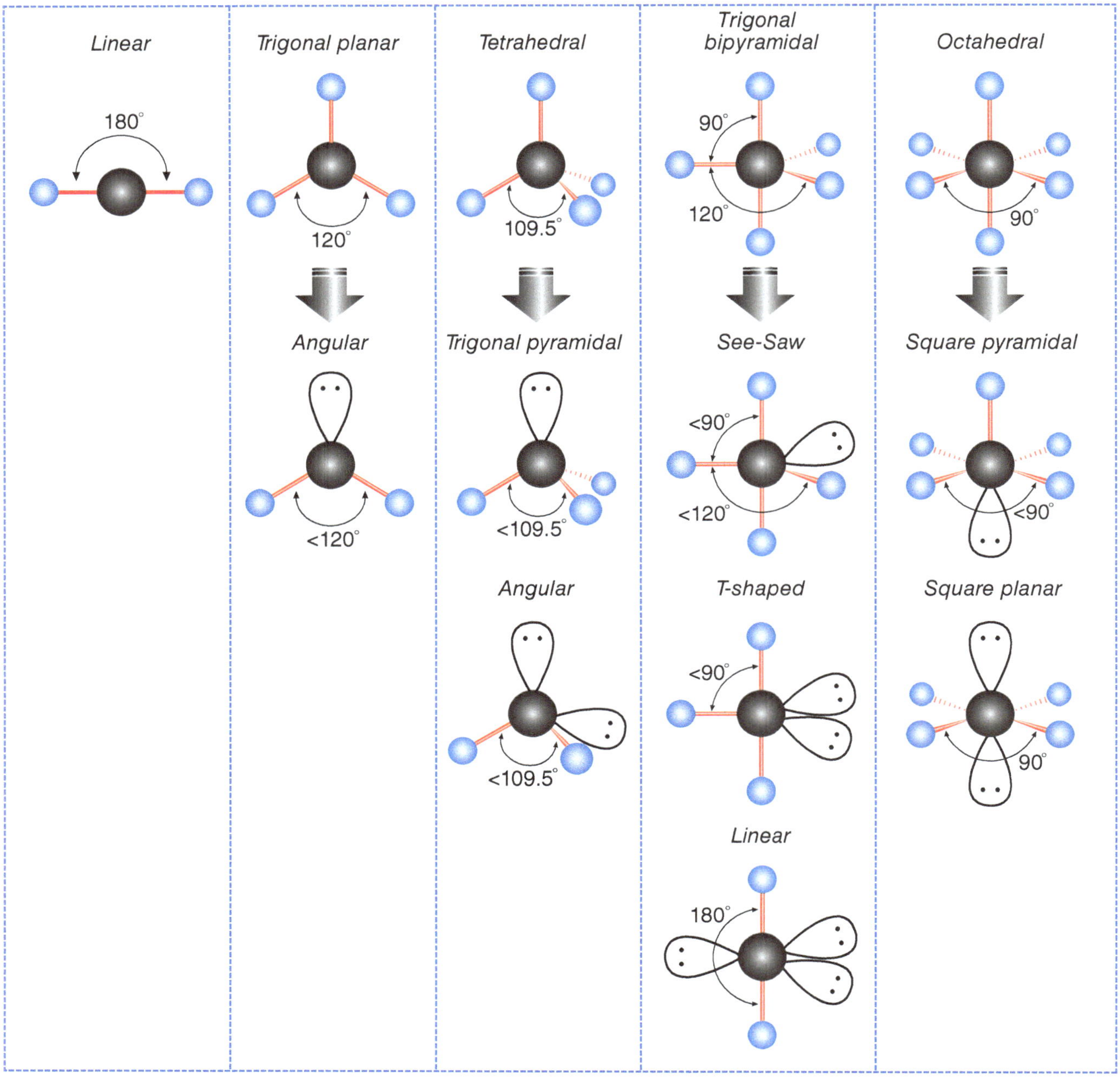

T-shaped molecule (<90° angle), and three lone pairs make a <u>Linear</u> molecule with 180°. See different ball-and-stick models in the table above. One lone pair in the octahedral structure makes a <u>square pyramidal</u> molecule and two make a <u>square planar</u> molecule. What is the best way to

My Professor Says:

draw Three-Dimensional (3D) Structures of molecules? We must know the parent electron group arrangement around the central atom and the bond angles predicted by VSEPR (VSEGR to be precise) — Valence Electron **Group** Repulsion— theory. Using the ball-and-Stick models helps us see how molecules look in 3D. To draw the 3D shapes, we use solid lines (—) to represent all bonds that lie in the plane of the paper, a dotted line (·····) for bonds that, would in 3D, extend to the back of the paper, and an elongated triangle with its apex attached to the central atom (◄) to represent any bond that, would in 3D, extend toward us from the plane of the paper. When drawing a 3D structure of any given molecule we must place as many atoms as possible in the plane of the paper. The position of lone pairs, only those around, a central atom may be shown with —, ----, or ◄ as described above because they affect the positions of bonded atoms. Since linear and trigonal planar arrangements are in the plane of paper their 3D structures consist of *only solid lines* drawn *180° apart for linear*, and *120° apart for trigonal planar*.

Two Linear- and one trigonal planar arrangement

To draw a tetrahedral, we first make 109.5° angle with two solid lines and bisect the angle. We then attach, to the center, a wedge and a dotted line along the bisecting line. Erasing the bisecting line leaves a tetrahedral arrangement of electrons.

Do I get it? Let me check:

Practice Questions

6.9. Count electron groups around each center and name *VSEGR-based molecular shape*.

Lewis Structure	No of Electron Groups per Center & Molecular Shape
[:Cl:—I(:Cl:)(:Cl:)—:Cl:]⁻	
:S: ‖ H—C—H	
[H—O—S(=O:)—O:]⁻	**e-groups:** 4 at **O** and 4 at **S**; **Shape:** Bent at **O**, trigonal pyramidal at **S**
:F:—As(:F:)(:F:)(:F:)—:F:	
H—C(H)(H)—C(:Br:)=C(H)—H	
:O:=Xe(:F:)(:F:)(:F:)(:F:)	

127

My Professor Says:

Do I get it? Let me check:

Steps 1, 2 and 3 of drawing a tetrahedral arrangement

In reality, a tetrahedral arrangement consists of two solid lines diverging 109.5° apart in one plane and a wedge and a dotted line also diverging 109.5° apart in an orthogonal plane, that is, the planes are *perpendicular* (90°) to each other.

The wedge and a dotted line appear adjacent to each other since the orthogonal plane is not in our line of view. Structures with multiple central atoms may have a combination of linear, trigonal planar and tetrahedral electron arrangement around the central atoms. For adjoining tetrahedral and trigonal planar centers we first draw a zig-zag line from a bonded atom to the first center, through the second center, and so on.

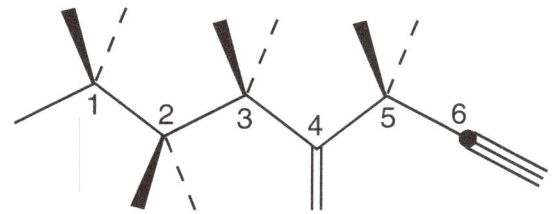

Tetrahedral at 1, 2, 3, 5; trigonal planar at 4, and linear at 6.

As we have seen with the ball-and-stick model on page 126, it is possible to have lone pairs on the central atoms. If there is one lone pair on center 2, then the geometric shape outlined by the atoms would be

Practice Questions

6.10. Draw the *VSEGR-based molecular shapes*.

Lewis Structure	VSEPR (3D) Structure
$[ICl_3]^-$ Lewis structure with I center bonded to three Cl atoms, with lone pairs	
H_2CS Lewis structure with C center double-bonded to S and single-bonded to two H	
$[HSO_3]^-$ Lewis structure with S center bonded to =O, -O:-, and -O-H	
AsF_5 Lewis structure with As center bonded to five F atoms	
$CH_3-CBr=CH_2$ (H-C-C=C-H with H, Br, H substituents)	
$XeOF_4$ Lewis structure with Xe center double-bonded to O and single-bonded to four F	

My Professor Says:

Do I get it? Let me check:

"Tetrahedral at 1, 3, 5; trigonal pyramidal at 2; trigonal planar at 4, and linear at 6." If there is one lone pair on center 2 and two lone pairs on center 3, then the geometric shape outlined by the atoms would be *"Tetrahedral at 1, 5; trigonal pyramidal at 2; bent at 3, trigonal planar at 4, and linear at 6."* If a central atom is an element in period 3, 4, or so on, then in addition to *s*- and *p*-orbitals, such a central atom has *d*-orbitals in its valence shell. Its orbitals can be mixed to create additional *hybridized* orbitals which can accommodate more than 4 pairs of electrons. We say, these elements exceed octet. Unlike Period 2 central atoms, which strive for noble gas configuration, these Period 3 elements strive to minimize their formal charge. A trigonal bipyramidal arrangement has three solid lines forming a T-junction as well as a wedge and a dotted line (in the orthogonal plane) from the central atom to the right. One can visualize an equatorial triangular plane with one solid line sticking up at 90° and another line downward also at 90°. Octahedral arrangement can be depicted as an equatorial square plane with one solid line sticking up at 90° and another line downward also at 90°. Making the square plane in this case, is the wedge extending from the page to us and a dotted line extending to the back while solid lines are in the plane of the page.

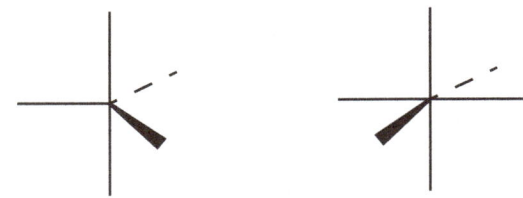

Trigonal bipyramidal   Octahedral shapes

My Professor Says:

Polarity of Molecules

A diatomic molecule (i.e., a molecule made up of only two atoms) is polar if the atoms have different electronegativity values. The larger the difference in electronegativity values the more polar the molecule. While polarity of diatomic molecules is directly related to bond polarity (usually indicated by dipole arrow), in molecules with three or more atoms, polarity of the molecule depends on the presence of a net-dipole. A net-dipole results from geometric imbalance in polarities of the bonds. *Any molecule with a parental shape and the same bonded atoms will be nonpolar.* Net dipole or imbalance in bond polarities results when atoms bonded to the central atom are different (or are having different electronegativity values). Consider tetrahedral molecules, for example. The molecules, CH_3Cl, CH_2Cl_2, and $CHCl_3$ are all polar while, CH_4 and CCl_4, are nonpolar.

Do I get it? Let me check:

Practice Question

6.11. Consider their 3D shapes and rewrite each of the following molecules in the appropriate column:

H_2O; CH_2F_2; NH_3; SO_3; SO_2; CH_4; BrF_5; CO_2; H_2S; PF_5; NF_3; BF_3

Polar	Nonpolar

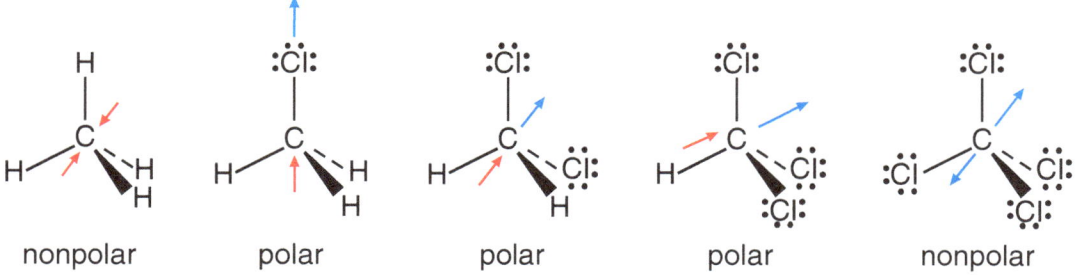

My Professor Says:

Molecules with one or more lone pairs around the central atom (these include Angular, Trigonal pyramidal, See-saw, T-shaped, and Square pyramidal) are usually polar even with the same bonded atoms because of asymmetry. In fact, asymmetry is the most important indicator of polarity in molecules, including those with multiple central atoms. In general, *asymmetric molecules are polar* while *symmetric molecules are nonpolar*.

Hybridization of Orbitals

The arrangement of electrons in a bonded atom, whether paired or unpaired, of a given element is that of an atom that underwent *hybridization*, that is, the mixing of orbitals to allow pairing of its electrons with those of other atoms with which it forms covalent bonds. Notice that valence electrons in a hybridized atom are arranged differently compared to those in an isolated atom. For example, the electron configuration of carbon is $1s^2\ 2s^2\ 2p^2$:

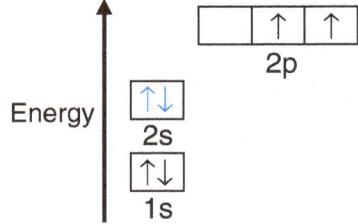

As this orbital energy diagram of an isolated carbon shows, not all 4 valence electrons are unpaired, only the two 2p electrons are. Yet carbon forms four covalent bonds in methane, CH_4. How?

Do I get it? Let me check:

Practice Question

6.12. Rewrite each of the following molecules in the appropriate column:

H_2O; CH_2F_2; NH_3; SO_3; SO_2; CH_4; BrF_5; CO_2; H_2S; PF_5; NF_3; BF_3

Polar	Nonpolar

131

My Professor Says:

The key is hybridization. Hybridization of carbon atom in CH_4 involves reshaping of its valence orbitals, 2s and 2p, into hybrid orbitals of equivalent energy, each called sp^3-orbital since it has mixed properties of the individual orbitals **s**, $\mathbf{p_x}$, $\mathbf{p_y}$, and $\mathbf{p_z}$. The four valence electrons are evenly spread in the new hybridized orbitals:

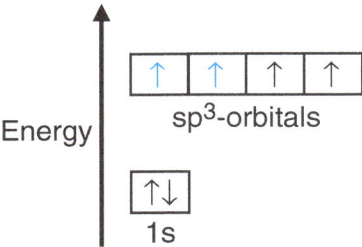

This way, carbon accommodates the four hydrogen atoms with their single electrons by forming covalent bonds. To assign the type of hybrid orbitals generated by the central atom we simply look at the Lewis structure of the compound. We designate electron groups (i.e., bonds—regardless of whether single, double, or triple—as well as lone pairs) around the central with the letters we use to name orbitals namely, **s** *(only 1 group)*, **p** *(up to 3 groups)*, **d** *(up to 5 groups)*, ... and identify the type of hybridization as **sp**, **sp^2**, **sp^3**, **sp^3d**, **sp^3d^2**, and so on depending on the number of electron groups around the central atom.

Do I get it? Let me check:

Practice Questions

6.13. Write the *hybridization(s)* of the central atom(s).

Lewis Structure	Hybridization of Central Atoms
$[ICl_4]^-$	
$H_3C-C\equiv N$	
$[HOSO_3]^-$	
AsF_5	
$H_3C-CBr=CH_2$	
$XeOF_4$	

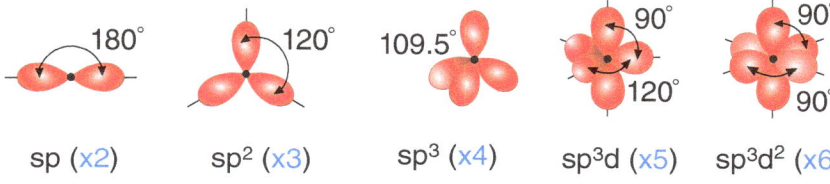

132

My Professor Says:

So, carbon in CH_4 is *sp³*-hybridized. Oxygen in H_2O is *sp³*-hybridized. Carbon in CO_2 is *sp*-hybridized.

Properties of a Covalent Bond

A covalent bond forms when two nonmetal atoms come close to each other such that the unpaired electron of one atom oscillates into the sphere of the other atom's nuclear pull. Notice that an unpaired electron generates magnetic field as it oscillates as does any moving charge. So, one atom's unpaired electron is magnetically attracted to the other atom's oscillating electron.

Once paired, the magnetic field is neutralized. Additionally, an atom has inherent tendency to attract electrons, called electronegativity. Among elements which form covalent bonds, fluorine is the most electronegative, followed by those radially closer to it. However, electron pairs repel each other electrostatically and maximize their radial distance from each other. The nuclei also repel each other. So, the balance between nucleus–electron attraction and nucleus–nucleus repulsion determines the length of the covalent bond. The *length* of a covalent bond is always shorter than the sum of atomic radii.

Do I get it? Let me check:

Practice Questions

6.14. Write the number of sigma bonds as well as pi bonds in each structure.

Lewis Structure	No. of Sigma bonds & pi bonds
$[ICl_4]^-$	σ: _____ π: _____
$H_3C-C\equiv N$	σ: _____ π: _____
$[HOSO_3]^-$ (HSO₄⁻)	σ: _____ π: _____
AsF_5	σ: _____ π: _____
$H_2C-CBr=CH_2$ (with H on first C)	σ: _____ π: _____
$OXeF_4$	σ: _____ π: _____

133

My Professor Says:

Do I get it? Let me check:

The difference between electronegativity values determines the *polarity* of a covalent bond. A single bond is always longer than a double bond, and a triple bond is the shortest. However, a triple bond is stronger than a double bond, and a single bond is the weakest. To determine the *strength* of a covalent bond, we calculate *bond order* which is the *number of bonding pairs divided by the number of bonded atoms*. For example, bond order in the structure of HCN molecule, **H-C≡N:** $= 4/2 = 2$.

Since there are only two groups around carbon in **H-C≡N:**, carbon is *sp*-hybridized as shown below.

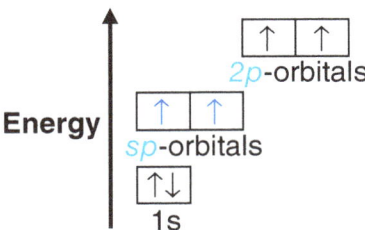

Nitrogen is also *sp*-hybridized. Hybridization is what elements do when they prepare to form bonds. It is similar to how young people prepare themselves to accommodate their special date. Once they are appropriately hybridized, the unpaired electrons in each orbital attract each other magnetically as we have described.

My Professor Says:

© Mr. Rashad/Shutterstock.com

Depending on their orientation in the *xyz*-planes, some orbitals, that is, electron oscillation regions will overlap *head-to-head making a sigma (σ) bond* while other orbitals will overlap *side-to-side making a pi (π) bond* as shown below for the **H-C≡N:** molecule.

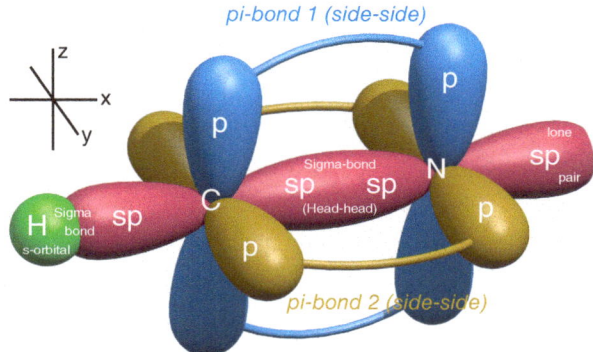

Modified from © izzmain/Shutterstock.com

So, the **H-C≡N:** molecule has two sigma (σ) bonds and two pi (π) bonds.

Do I get it? Let me check:

© AVIcon/Shutterstock.com

Practice Questions

6.15. What type(s) of orbital overlaps are there in the nitrogen molecule (:N≡N:)? Choose.

a) Three head-to-head (σ) overlaps only

b) Three side-to-side (π) overlaps only

c) Two head-to-head (σ) overlaps, one side-to-side (π) overlap

d) Three head-to-head (σ) overlaps and two side-to-side (π) overlaps

e) One head-to-head (σ) overlap, one side-to-side (π) overlap

6.16. Hybridizations of the central atoms in the structure of acetic acid, **CH₃(CO)OH**, (shown in bold letters), listed in order from left to right, are:

a) sp^3, sp, sp b) sp^3, sp^2, sp^3

c) sp^3, sp^3, sp^3 d) sp^3, sp^3, sp

e) sp^3, sp^2, sp

6.17. How many sigma (σ) bonds can be formed by an sp² hybridized central atom? Choose.

a) 2   b) 3   c) 4   d) 5   e) 6

135

THESE ARE MY OWN NOTES
(from listening to lectures, watching YouTube videos, etc.)

Date: _____

Date: _____

Date: _____

Date: _____

I wish the professor could explain this: _____

Molecular Structures

Molecular Structures

Molecular Structures

GASES AND CONDENSED PHASES

The Behavior of Gases

The *volume*, **V**, of a gas depends on the *pressure*, **P**, and *temperature*, **T**, of that gas. Pressure in turn depends on the number of gas particles or amount of gas (usually in *moles*, **n**). These are the four physical parameters we focus on to describe and predict the behavior of gases. In general, gas behavior is understood in terms of the kinetic molecular theory which states that at constant temperature, each gas particle moves in a straight line and may change its direction after colliding with the container wall or with another gas particle, but it does not lose or gain kinetic energy.

Individual Gas Laws

We change the quantity of one parameter and observe how the other parameter responds while the other two parameters are constant. For example, we record measurements of **P** *versus* **V** with T & n constant, **T** *versus* **V** with P & n constant, or **P** *versus* **n** with T & V constant. We then plot values of the manipulated parameter (as x-values) and those of the responding parameter (as y-values) on a graph. Except for volume (V) *versus* pressure (P) which are inversely related, all other sets of two parameters are directly proportional to each other, that is, they change

Practice Questions

7.1. Label the following phases (or states of matter)

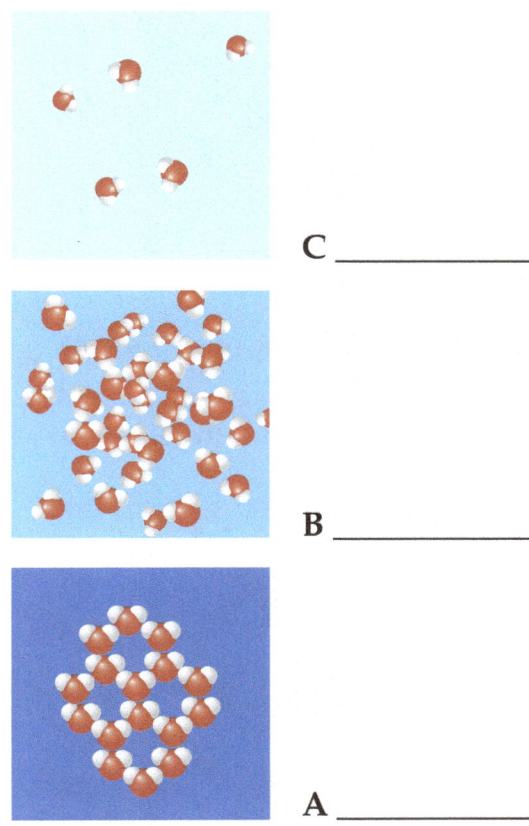

© OSweetNature/Shutterstock.com

7.2. Write the chemical formula of this substance. _____

7.3. If these are at the same temperature, **T**, and pressure, **P**, which particles have the *highest energy*?
_____

143

| My Professor Says:  | Do I get it? Let me check:  |

by the same factor. We can plot pressure (**P**) *versus* inverse of volume ($1/V$) to get a linear graph of the form "$y = mx + b$." This equation can be written explicitly as:

Responding = (Const. × Manipulated variable) + y-intercept

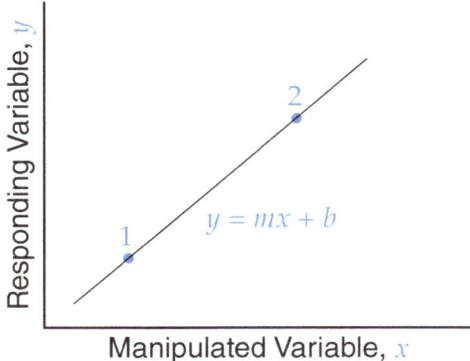

The usable equations (shown in boldface below) are called *Gas Laws* and have been named after the scientists who derived them originally:

1. Robert <u>Boyle</u>: $\boldsymbol{P \propto \dfrac{1}{V}}$ or $\boldsymbol{PV = }$ *constant*

   So, $\boldsymbol{P_1 V_1 = P_2 V_2}$

2. Jacques <u>Charles</u>: $\boldsymbol{V \propto T}$ or $\dfrac{V}{T} =$ *constant*

   So, $\dfrac{V_1}{T_1} = \dfrac{V_2}{T_2}$

3. Joseph <u>Gay-Lussac</u>: $\boldsymbol{P \propto T}$ or $\dfrac{P}{T} =$ *constant*

   So, $\dfrac{P_1}{T_1} = \dfrac{P_2}{T_2}$

4. Amedeo <u>Avogadro</u>: $\boldsymbol{V \propto T}$ or $\dfrac{V}{n} =$ *constant*

   So, $\dfrac{V_1}{n_1} = \dfrac{V_2}{n_2}$

7.4. Which particles make the *least condensed* phase? _____

7.5. Which picture has the *highest number* of moles, **n**? _____

My Professor Says:

Do I get it? Let me check:

Practice Questions

7.6. A gas is shown in a 4.0 L container at 25°C.

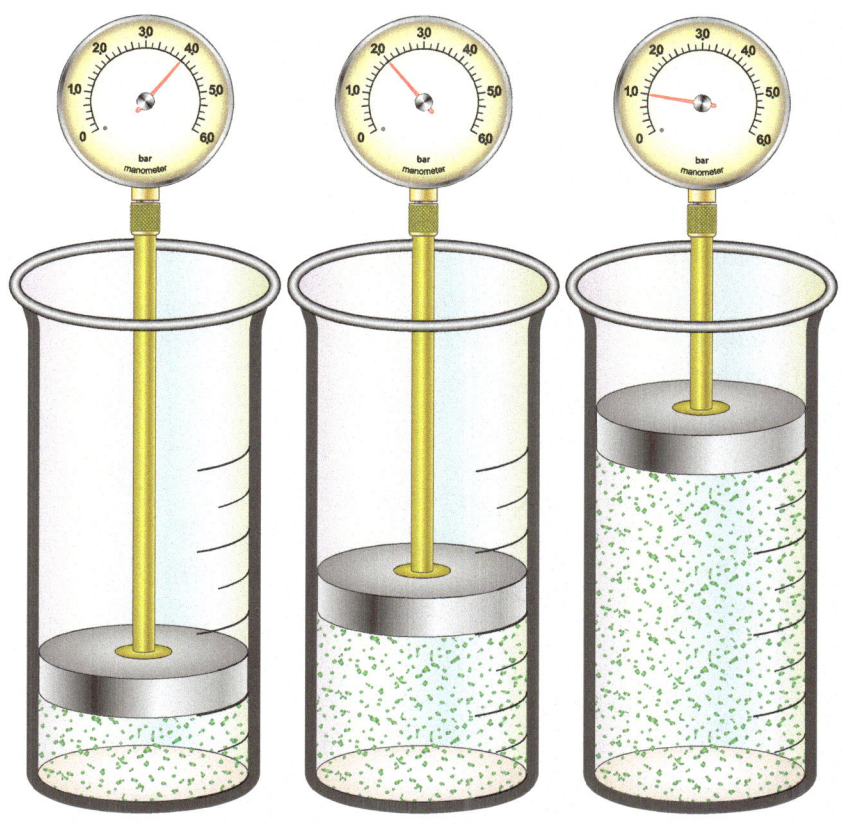

The Combined Gas Law

In fact, the relationship discovered by Avogadro lead to the calculation (by others) of the number of gas particles in one mole, 6.022×10^{23}, which was named Avogadro's number in his honor after the charge of one electron (1.602×10^{-19} C by Robert Millikan) as well as charge of one mole of electrons (96485 C by Michael Faraday) were determined. You see, Avogadro's number is simply Faraday's constant

a) Which gas law is demonstrated?
 _____

b) Which picture shows the highest pressure? _____.

c) In this experiment (left to right) which gas property is manipulated? _____. Which one is responding?
 _____

145

My Professor Says:

divided by charge of one electron. The usable equations for changing conditions can be combined into one law called the *combined gas law* as follows:

$$\frac{P_1 V_1}{n_1 T_1} = \frac{P_2 V_2}{n_2 T_2}$$

The Ideal Gas Law

We can also combine Boyle's, Charles', and Avogadro's linear relationships into one law called the *ideal gas law*. Notably, the Gay-Lussac relationship is maintained in the ideal gas law:

$$V \propto \frac{1}{P} \times nT \text{ or } P \propto \frac{1}{V} \times nT \text{ or } PV \propto nT$$

So, $\frac{PV}{nT} = Gas\,Constant, R$

The gas constant, $R = \frac{0.0821\,atm.L}{mol.K}$. At exactly 273 K (or 0°C) and 1 atm, called Standard Temperature and Pressure (**STP**), one mole of the ideal gas is found to be 22.4 L. This is referred to as standard molar volume of gas. The extent to which real gases display linearity as predicted mathematically by the gas laws is called the *ideal gas behavior*. The ideal gas law is commonly written as **PV = nRT**. Note that the *ideal gas* is hypothetical. Real gas particles do deviate from ideal behavior because they experience forces of attraction. Note some helpful tips

Do I get it? Let me check:

d) Is pressure, **P**, of a gas *directly* or *inversely* related to volume, **V**?
_____

How can we find out? _____
_____

e) Find the pressure of 3.5 L of the gas.

f) What temperature will change gas volume from 2.0 L to 3.5 L at 2.0 bar?

My Professor Says:

which you can use when solving gas law problems:

1. Different units of pressure can be converted. [1 atm = 14.7 psi = 760 mmHg = 760 torr = 1.010325 bar = 101.325 kPa]

2. Always use temperature in units of Kelvins (K) and volume in units of liters (L).

3. For changing conditions where two parameters were measured, use the applicable individual laws. If unsure of the law to use, you may simply use the combined gas law, but you must eliminate (cancel out) parameters that are constant.

   a) For changing conditions where all four parameters have been measured, use the combined gas law.

   b) For nonchanging conditions where three parameters have been measured, use the ideal gas law.

   c) For finding molar mass (M_r) of an unknown gas where the mass (m) is given, use the rearranged form of the ideal gas law:

$$M_r = \frac{mRT}{VP}$$

Do I get it? Let me check:

Practice Questions

7.7. The volume of this basketball is 434 cm³ and its *gauge* pressure is 7.63 psi at 25°C.

a) The air pressure inside is higher than in the atmosphere (14.7 psi) outside. Find *true* pressure of air inside this ball? *If it is higher, I can _____ atmospheric pressure to gauge pressure. I can do this:*

b) How many moles of air is in this ball? *I need the ideal gas law and units that can cancel those of R. Let me try:*

c) What mass of air (M_r = 28.97 g/mol) do we need to make its pressure 8.4 psi?

147

My Professor Says:

Do I get it? Let me check:

d) For finding density (D) which is given by m/V of an unknown gas where the molar mass is given, use the rearranged form of the ideal gas law:

$$\frac{M_r P}{RT} = \frac{m}{V} = D$$

e) For finding one of the four parameters where the mass of a reactant and a balanced equation is given, first use dimension analysis (stoichiometry) to find moles of the gaseous product and then use the ideal gas law.

The Law of Partial Pressures

John Dalton discovered that a gas mixture has a total pressure, P_T, and each of the component gases have a partial pressure P_1, P_2, and so on. The partial pressure of a component gas is given by the mole fraction (χ) of the gas multiplied by the total pressure:

$$P_1 = \chi\, P_T$$

where $\chi_1 = \dfrac{mol\ Gas\ 1}{(mol\ Gas\ 1\, +\, mol\ Gas\ 2\, +\, ...)}$

This law helps us understand that in a given container of a fixed volume, particles of a given gas will distribute themselves in the same way they would if other gases were not present. Unless the component gas reacts chemically, or the container volume changes, its partial pressure will not be changed by the addition of other gasses.

Practice Questions

7.8. A weather balloon temperature changes from 30°C to 50°C. Its final volume, V_2, is twice the original volume, V_1. Find its final pressure, P_2, relative to initial pressure, P_1

7.9. Sodium azide in car airbags decomposes upon impact at 25°C:

$$2\ NaN_3(s) \rightarrow 2\ Na(s) + 3\ N_2(g)$$

[Note: Na produced is stabilized by other ingredients]

What mass of NaN_3 will inflate a 46.6 L airbag of a car to a pressure of 1.12 atm?

7.10. Air in human lungs contains 0.062 mol $H_2O(g)$, 0.74 mol $N_2(g)$, 0.050 mol $CO_2(g)$, and 0.12 mol $O_2(g)$ at total pressure 1.8 atm. Find partial pressure of oxygen, $O_2(g)$.

THESE ARE MY OWN NOTES
(from listening to lectures, watching YouTube videos, etc.)

Date: _____

_____
_____
_____
_____

Date: _____

_____
_____
_____
_____
_____
_____

Date: _____

_____
_____
_____
_____
_____
_____

Date: _____

_____
_____
_____
_____
_____

I wish the professor could explain this: _____

_____
_____
_____
_____
_____

Student's Practical Laboratory Experiment
DISCOVERING PHYSICAL LAWS OF GASES

OBJECTIVE

The objective of this experiment is to measure pressure, volume, temperature, and amount of gas, and plot the results in order to determine the relationships between these gas properties. In this lab exercise **air** will be used to demonstrate the behavior of gases in terms of the four gas properties mentioned. We will use the apparatus shown below or their variations.

© Sentavio/Shutterstock.com   © udaix/Shutterstock.com   © BW Folsom/Shutterstock.com

LEARNING OUTCOMES

After this experiment, students will be able to:

1. *Measure volume of air at various levels of pressure*
2. *Use a computer software, for example, MS Excel spreadsheet* to plot a graph.
3. *Describe the relationship* between volume, as well as amount, and pressure of a gas

SAFETY NOTES

Do not exceed 50 psi (345 kPa) when pumping air into the rigid plastic bottle. Always wear safety goggles and shoes when doing experiments. *When finished, wash your hands with soap and water.*

Materials Needed:

1. A 1.06 L rigid plastic bottle with stem valve—to increase gas pressure in a fixed volume.
2. A small calibrated syringe with cap—to be placed in a pressurized plastic bottle
3. A hand pump or air compressor—to pressurize a container to 50 psi
4. A tire gauge—to measure gas pressure inside the bottle
5. A thermometer—to measure temperature of air in the room
6. A balance—to measure mass

PRE-LAB ASSIGNMENT

Follow the steps below to plot a graph of Volume of a gas *(as y)* **versus** Pressure *(as x)*.

1. Copy the table on to an MS Excel spreadsheet.
2. Highlight the first two columns and click "insert" at the top menu bar of Excel, then "XY scatter" to plot the data. Notice, this graph is not linear!
3. Hover over the top right corner of the chart and click "+" to add titles.

Volume, V (mL)	Pressure, P (kPa)	P × V	$\dfrac{1}{\text{Volume}}$
34.0	100		
23.0	150		
17.0	200		
13.8	250		
11.3	300		
9.7	350		
8.5	400		
7.6	450		
6.8	500		
6.2	550		

4. Notice that at any point along the curve the product *PV* is constant (to the nearest 100).
5. A more meaningful way to graph this data, from which we will see a linear relationship, might be $P = k/V$, or $P = k\left(1/V\right)$, which is a straight line. Use MS Excel (or equivalent) to complete the data table, and then plot the graph of **pressure, P** *(as y)* vs. **volume inverse,** $1/V$, *(as x)*.

   a) Move the mouse over one of the data points, and right-click, then choose "add trendline" on the window that appears; select "linear," and check "set intercept to 0.0," "display equation," and "display correlation, r²."

   b) Move the mouse over the top right corner of the chart and click "+" to add titles.

6. Display the following information on your graphs:

   a) Title of the graph: "Boyle's Law for Gases"

   b) The y-axis, "Pressure," and x-axis, "Volume (mL)" on the initial curve only and on the linear graph label x-axis as "$1/v$." Thus, y- and x-axes must be labeled according to the titles of the columns in the data table.

   c) Equation of the linear graph, as well as the correlation factor (r^2 value) for the linear graph.

7. Save your spreadsheet:

   a) Right-click twice on the tab "Sheet 1" at the bottom of the spreadsheet, select rename, and change it to "Gases Pre-Lab."

   b) Click File at the top menu, then save as [*filename here*.xls]. You will add other spreadsheets to this file. So, remember the filename. Submit your pre-lab electronically to your instructor.

EXPERIMENTAL PROCEDURE

Useful tips for this experiment:

- Write the units of measurement, either in the column title or next to each value.
- You need 10 data points or more for each experiment.
- Do not deflate between measurements since air leaks during each measurement.
- Be swift when taking pressure measurements to minimize loss of air.
- The gauge only measures pressure higher than atmospheric pressure.

Experiment 1

[Pressure (*P*) <u>vs.</u> mass (*m*) at constant *V* and *T*].

1. For this experiment, you will use the compressor or air pump to pressurize air into the rigid plastic bottle to 40 psi maximum. The container will keep the volume constant, but as you increase pressure, the bottle gets warmer! *Wait for the bottle to cool to room temperature.*
2. We need mass of air alone, and to get it we must know mass of the empty bottle. Realize that when you start the experiment your "empty" container is not empty—it is full of air at atmospheric pressure. So, we cannot find the "empty" weight of the container by experiment. *We will do this with our graphical analysis.*
3. Go ahead and weigh the pressurized bottle. Always measure the mass first followed by the pressure (in kPa) because, as you will notice, each pressure measurement changes the amount of mass in the bottle. For this reason, there is no need to reduce the pressure between measurements. Simply weigh and take the next pressure measurement.
4. Record your readings in the data table below.

Experiment 1 Data.

Gauge Pressure (kPa)	Total Mass (g)	Gauge Pressure (kPa)	Total Mass (g)	Gauge Pressure (kPa)	Total Mass (g)	Gauge Pressure (kPa)	Total Mass (g)

Instructor's OK _____

Experiment 2

[Pressure (**P**) vs. Volume (**V**) of fixed moles of gas, **n**, at constant **T**].

1. For this experiment, you will fill a small calibrated syringe with air at atmospheric pressure, cap it, and put it inside the rigid plastic bottle. Cut the flanges so that it fits into the bottle.
2. Now use the compressor or pump to pressurize air into the rigid plastic bottle to 40 psi maximum.
3. Read the volume of air in the syringe through the side of the clear container before you measure pressure (in kPa) exerted on the syringe.
4. Go ahead and take the next volume reading followed by pressure measurement.
5. Record your readings in the data table below.

Experiment 2 Data.

Volume (mL)	Gauge Pressure (kPa)	Volume (mL)	Gauge Pressure (kPa)	Volume (mL)	Gauge Pressure (kPa)	Volume (mL)	Gauge Pressure (kPa)

Instructor's OK _____

EXPERIMENTAL RESULTS

Copy your data into MS Excel (or equivalent) spreadsheet and plot appropriate graphs to find mathematical relationships for the variables. You need at least 10 well-spaced data points. For conversion of units please note the following:

- The molar mass of air = 28.97 g/mol
- Atmospheric pressure = 101 kPa or 14.7 psi
- Total mass = bottle + pressurized air
- Gauge pressure = pressurized air − atmospheric pressure

DATA ANALYSIS:

Follow the instructions below to navigate MS Excel, and to analyze your data so that you can find mathematical relationships between the variables.

1. On your electronic device, open **MS Excel** and find the file in which you saved your pre-lab. Click on the tab "Sheet 2" or "+" at the bottom of the spreadsheet to open the next spreadsheet.

    a) Start with Experiment 1 data (**P** vs. **m**).

    b) Make your data table:

    i) Type in the column titles, "Corrected pressure (kPa)" in column A row 4 (i.e., cell A4) and type "Gauge pressure (kPa)" in cell B4, "Total mass (g)" in cell C4.

    ii) To have words wrapped in one cell, select "wrap text" on the top menu bar of the "home" screen.

    iii) Type the data you have, accurately in appropriate columns!

    iv) Notice you do not have data for column A, "Corrected pressure (kPa) ." Click on cell A5 and type: = **B5 + 101 enter**. Now click cell A5 again and move the mouse to find a "+," at the bottom right corner of the cell, and drag down to the end of your columns. This will copy the formula for each data point.

    c) Plot the graphs:

    i) Highlight the two columns (B_:C_) of data.

    ii) Click "Insert" on the top menu bar of the home screen. Then find "Charts" and select "Scatter XY." This will plot the graph of "corrected Pressure (y-axis)" vs. "Tot. *mass* (x-axis)."

    iii) You will notice that the graph of your data is **linear**. So, you do not need to search for a relationship, it is already in $y = mx + b$ form. This demonstrates that pressure of air varies linearly with the "total mass."

    iv) Right-click on any data point and on the window that appears select "Add Trendline," then select "Linear" to insert a linear trendline.

    v) Then check appropriate boxes at the bottom to display the straight-line equation ("y = mx + b" format) and a correlation factor (r^2). Keep this graph.

    vi) However, **your units on the x-axis are kPa**. We need **atm**. Use $\frac{1 \text{ atm}}{101.325 \text{ kpa}}$ as a conversion factor so that kPa cancels out. This is the same as dividing each kPa value by 101.325 kPa.

    vii) Type "Corr. Pressure (atm)" in cell D4. Click on cell D5 and type: = **A5/101.325 enter**. Find a "+," at the bottom right corner of cell D5, and drag down to the end of your columns.

viii) **Are units on the y-axis correct?** Again, they are not. Our data is in **grams**. We need **moles**. Use the molar mass of air (28.97 g/mol) to convert units. But before we begin, notice that the graph of "Tot. *m*" vs. "corrected **P** (kPa)" has a **y**-intercept. *Shouldn't zero pressure correspond to zero grams of gas? Oh yeah. We used total mass (air + container). So, the y-intercept is the mass of the container alone. We should subtract it out from our data before we convert it into moles of air.*

ix) Type "Mass of Air (g)" in cell E4, and "Moles of Air (mol)" in cell F4. Click on cell E5 and type: = **C5** – *y-intercept here* **enter**, from previous graph, [see step C(v) above]. Click on cell F5 and type: = **E5/28.97 enter**. Highlight cells E5 and F5. Then find a "+," at the bottom right corner of cell F5, and drag down to the end of your columns.

x) Since **P**(atm) and moles (**n**) of air columns are not adjacent, hold down Ctrl key as you highlight them. Repeat steps C(ii) – C(v) above to plot a new graph of "corrected **P** (atm)" *versus* "**m** of air(g)."

xi) Finally, your data is in good shape with correct units, and no extra or missing information. You have linear data, the line goes through the origin (or y intercept is close to zero). The linear fit is reasonable (i.e., r^2 is close to 1). The value in front of the x in the equation on the graph is the slope of your line. It is the *proportionality constant* between your two variables, and has units of *moles/atm*.

xii) Right-click twice on the tab "Sheet 2" or "+" at the bottom of the spreadsheet, select rename, and change it to "Experiment 1."

2. While you are still on **MS Excel** program, click on "sheet 3" or "+" also at the bottom to open the next spreadsheet for Experiment 2. *Now follow the same steps on your own*.

    a) Type in your data and build a graph in its original form.

    b) Look at what we did before and ask yourself the same questions.

       i. Correct units first.

       ii. Find a direct, linear relationship.

       iii. Re-graph your data so that all units are correct and it is linear. Find the equation that describes the relationship and the proportionality constant (or slope).

3. OPTIONAL: You may request "Experiment 3, V vs. T" data from your instructor for extra points.
4. Save the entire spreadsheet and e-mail it to yourself.
5. Now that you have analyzed your data, follow the outline on the next page to **write a lab report** on MS Word (or equivalent).

    a) Format your Excel graphs. Then copy and paste them into the final report.

    b) Copy the data tables you filled out during actual lab experiment.

    c) Make sure you answer the questions contained in the report.

    d) Save your final lab report and submit it electronically to your instructor.

BEHAVIOR OF GASES

LAB REPORT

[Follow the outline below and use MS Word to write a full report and submit it electronically for grading.]

****

Title:	[a simple name or phrase for the work]
Abstract:	[Summarize (in 50 words or less) the gas relationships you found.]
Introduction:	[Write a brief statement (50–100 words) of why this experiment is important. As background, use the *kinetic molecular theory* to explain the behavior of gas particles.]
Materials:	[List everything you used; give exact description, sizes, or quantities.]
Procedure:	[Describe (in your own 200–300 words) how you did the work so it can be repeated. Give main steps.]
Results:	[Insert your data tables and graphs with captions. Discuss whether your observations and data support the *kinetic molecular theory*. Assuming air temperature of 23°C, explain or show how to find the universal gas constant, R, of the ideal gas law $PV = nRT$ (in units of $atm \cdot L \cdot mol^{-1} \cdot K^{-1}$)]
Conclusions:	[In less than 50 words of your own, state what the results imply or what the graphs show, possible improvements, and how this information can be useful.]

[A general lab report rubric shown on the next page will be used to grade final lab reports.]

ASSESSMENT RUBRIC FOR GENERAL CHEMISTRY LAB REPORTS [20PTS MAX]

Demonstrate Scientific Method: Observation, Hypothesis, Experiment, Theory (or Law) statement

Learning Outcomes. Show that you can:	Ratings: Your work will be graded as			
	Excellent (A) if you can:	**Good (B)** if you can:	**Average (C)** if you can:	**Improving (D)** if you can:
1. Understand important science concepts	correctly use two or more science concepts to explain what happens **[5 pts]**	use one of the science concepts to explain what happens **[4 pts]**	correctly define important concepts **[3.5 pts]**	determine which concepts are described or defined **[2.5 pts]**
2. Apply the scientific method	perform all steps of the scientific method correctly **[5 pts]**	perform some steps of the scientific method correctly **[4 pts]**	recognize all the steps of scientific method and perform some when guided **[3.5 pts]**	recognize a few steps of the scientific method but not when they are used **[2.5 pts]**
3. Use math tools	determine what data to collect, graph it, and interpret it or do necessary calculations **[5 pts]**	collect data, graph it, and interpret it or do calculations with guidance **[4 pts]**	collect, graph, interpret and/or perform calculations when guided **[3.5 pts]**	collect data and try to graph it but need help to interpret it or to perform calculations **[2.5 pts]**
4. Be understood	write lab reports that are clear, or keep a lab notebook organized **[5 pts]**	write lab reports with minimum mistakes, or make minimum errors in lab notebook **[4 pts]**	write lab reports even if with significant mistakes, or keep lab notebook even if in a wrong format **[3.5 pts]**	write some parts of a lab report, or try to organize a lab notebook **[2.5 pts]**

--oOo--

My Professor Says:

Do I get it? Let me check:

Intermolecular Forces

Intermolecular forces are forces that hold liquid particles together. They arise from the attraction between one molecule and another molecule, or between ions and solvent molecules. Intermolecular forces are relatively **weak** compared to (ionic or covalent) bonding forces because they involve smaller partial charges that are farther apart. *Intermolecular forces* are responsible for nonideal behavior of gases and for existence of condensed states of matter. They are responsible for bulk (or colligative) properties of matter such as boiling points and melting points. There are three *major* types of intermolecular forces. In order of decreasing strength, they are:

1. *Hydrogen bonding* (between polar molecules that have at least one **F–H, N–H,** or **O–H** bond in their Lewis structures)

2. *Dipole–dipole forces* (between polar molecules)

3. *London (Dispersion) forces* (between nonpolar molecules)

Ions attract several polar solvent molecules at a time resulting in *Ion–Dipole interactions* which are stronger than hydrogen bonding. The term used is *Interparticle forces* (IPFs). We will discuss this further under solution formation.

The conditions under which a given pure substance exists as solid, liquid, or gas is shown by a phase diagram in which pressure is plotted against temperature.

Practice Questions

7.11. State the strongest *type of intermolecular forces* that occur between the following in the liquid phase:

a) $\ddot{\text{S}}=\text{C}=\ddot{\text{S}}$ and SO_3

b) NH_3 and SO_2

c) CH_2Br (CH₃Br) and PF_5

161

My Professor Says:

A typical phase diagram is shown below.

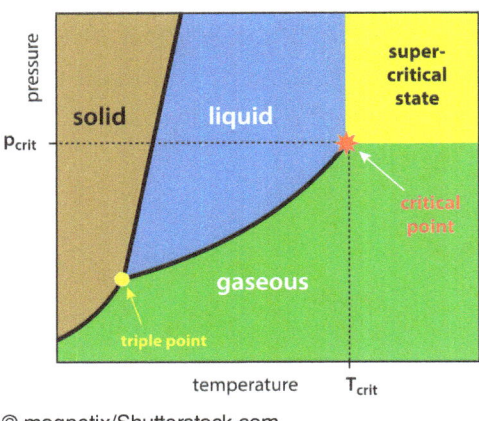

Each boundary indicates phase change. Any point along the boundary between solid and liquid, that is, the fusion curve, is a melting/freezing point; any point along the boundary between liquid and gas, that is, the vaporization curve, is a boiling/condensation point; and any point along the boundary between solid and gas, that is, the sublimation curve, is a sublimation/deposition point.

Solution Formation

A solution results from the following steps:

1. Solute particles separate from each other, while the solvent molecules also separate from each other. Each process is **endothermic**.

2. Solute and solvent particles attract each other. This process is **exothermic**.

The solution formation process is *endothermic if* the amount of *energy absorbed* in <u>step 1</u> is *larger than* the amount of *energy*

Do I get it? Let me check:

Practice Questions

7.12. The critical point of substance X is 304 K and 72.9 atm

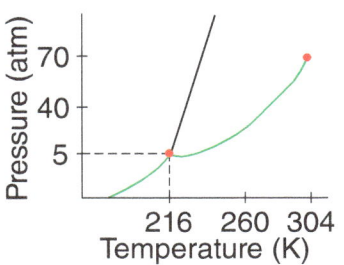

a) According to this phase diagram, what is the physical state of substance X at 760 torr and 25°C?

_____

b) What phase change will occur to substance X if it is initially at 260 K and 45 atm and then temperature is decreased to 210 K while keeping the pressure constant?

_____

released in *step 2*. Solution formation is *exothermic* *if* the amount of *energy released* in *step 2* is larger *than* the amount of *energy absorbed* in *step 1*.

$$\Delta H_{soln} = (\Delta H_{solute-solute} + \Delta H_{solv.-solv.})_{separ.} + (\Delta H_{solute-solv.})_{attr.}$$

Colligative Properties

Colligative properties are properties that we observe because of the presence of a large *number* of particles, regardless of chemical identity. Colligative properties are affected by the presence of solute particles. So, the four major colligative properties are best listed as:

1. *Boiling point elevation (ΔT_b)*. Boiling temperature is *elevated* because solute–solvent particle attractions are stronger than solvent–solvent attractions in pure solvent.

2. *Freezing point depression (ΔT_f)*. Freezing temperature is *depressed (or lowered)* because solute–solvent particles attract each other more strongly than solvent–solvent attraction. So, salty water melts below the freezing point of pure water.

3. *Vapor pressure lowering (ΔP)*. Since solvent–solvent particle attractions are normally weaker than solute–solvent attractions, molecules of a pure solvent go into the vapor phase more easily than when solute particles are present. According to Raoult's law, when a solute is dissolved, the vapor pressure of pure solvent (ΔP^o) is *lowered*.

Practice Questions

7.13. A semipermeable membrane separates a nonelectrolyte sugar, threitol ($C_4H_{10}O_4$) in solution, from pure water at 27°C.

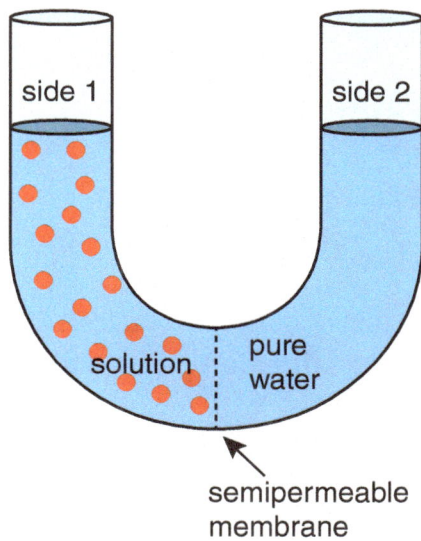

semipermeable membrane

a) Will the height of the solution (side 1) remain equal to that of pure water (side 2) over time?

_____

Explain:
_____
_____

b) What will likely happen if the tube is exposed to freezing temperature, 0°C?

_____

Explain.
_____
_____

My Professor Says:

4. *Osmotic pressure* (Π). When there is high concentration of solute particles in one compartment than in another connected to it which contains pure solvent, more solvent molecules migrate toward the concentrated compartment to dilute it, exerting pressure.

The applicable equations are:

i. $\Delta T_b = i\, m\, k_b$
ii. $\Delta T_f = i\, m\, k_f$
iii. $\Pi = \dfrac{i(MRT)}{V}$

[*i* is number of solute particles per formula unit; *m* is molality; and *M* is molarity of the solution.]

iv. $P_{solution} = \chi_{solvent}\, P^o_{solvent}$
v. $\Delta P_{solvent} = i(\chi_{solute}\, P^o_{solvent})$
vi. $P_{Total} = (\chi_{Liq\,A}\, P^o_{Liq\,A}) + (\chi_{Liq.\,B}\, P^o_{Liq\,B})$

[χ is the mole fraction, and P^o is the vapor pressure of a pure solvent.]

Do I get it? Let me check:

c) Find *molarity* (M) of threitol solution that causes osmotic pressure of 9.6 atm.

d) Find vapor pressure of 1.84 mol threitol solution (side 1) in 27.7 mol water [ΔP° (H$_2$O) = 26.7 torr]

Practice Questions

7.14. Consider these aqueous solutions [$m \approx M$]: 0.01m MgCl$_2$; 0.01m KCl; 0.01m C$_{12}$H$_{22}$O$_{11}$ Select the solution that matches the colligative property listed below:

a) Highest freezing point _____
_____

b) Lowest freezing point _____
_____

c) Highest boiling point _____
_____

d) Lowest boiling point _____
_____

e) Highest osmotic pressure _____
_____

My Professor Says:

© Mr. Rashad/Shutterstock.com

Practice Questions

7.15. What is the vapor pressure of a solution of 25.8 g urea (CH_4N_2O), a nonelectrolyte, dissolved in 275 g water at 45°C.

[$P^o(H_2O)$ at 45 °C is 71.9 torr]

Do I get it? Let me check:

© AVIcon/Shutterstock.com

7.16. If 0.0500 mol. CH_2Cl_2(l) and 0.0500 mol. CH_2Br_2(l) are mixed at 25°C, which component will be more in the vapor? Show your work.

[$P^o(CH_2Cl_2) = 133$ torr; $P^o(CH_2Br_2) = 11.4$ torr]

_____ because: _____

_____

_____

THESE ARE MY OWN NOTES

(from listening to lectures, watching YouTube videos, etc.)

Date: _____

Date: _____

Date: _____

Date: _____

I wish the professor could explain this: _____

Student's Practical Laboratory Experiment

FORCES THAT HOLD MOLECULES

OBJECTIVE

The objective of this experiment is to demonstrate that intermolecular forces determine physical properties of a substance. Each physical state of matter (i.e., gas, liquid, or solid) is a phase, a physically distinct, homogeneous system. The physical properties of each phase of matter are determined by the balance between the potential and kinetic energy of its component particles. Potential energy, associated with attractive forces, holds particles together while kinetic energy, associated with movement, tends to disperse the particles. It is also important to note that chemical behavior of each phase of matter is the same as that of the individual particles that make it up.

We can consider the bonds within a molecule as *intra*molecular forces. Between different molecules such as in a liquid or molecular solid, there are nonbonding attractions called *intermolecular forces* (IMFs) which hold them together as a substance. The strengths of these forces differ from one physical state to another and give rise to differences in physical behavior between phases of matter. When a liquid substance boils, only the *inter*molecular forces are broken, not the individual molecules. Major types of intermolecular forces in pure liquid substances include *Hydrogen bonding*, *Dipole–Dipole forces,* and *London dispersion forces.*

LEARNING OUTCOMES

After this experiment, students will be able to:

1. Identify the substance with relatively strong intermolecular forces.
2. Relate intermolecular forces to the structure of individual molecules.
3. Distinguish bulk properties from those of individual molecules.

SAFETY NOTES

Handle glassware with care and inspect it for cracks and sharp edges before use. Work in a fume hood or close to an air vent to avoid inhaling fumes from the liquids. Immediately wash off the chemicals should you make skin-contact. Always wear safety goggles and shoes when doing experiments. <u>When finished, wash your hands with soap and water.</u>

PRE-LAB ASSIGNMENT

1. Draw the **a)** Lewis and **b)** VSEPR-based structures for each formula. Fill in the blanks:

   Water H_2O Molar Mass: _____

                                                                   Polarity: _____

   n-hexane $CH_3CH_2CH_2CH_2CH_2CH_3$ Molar Mass: _____

                                                                    Polarity: _____

   2-Propanol $CH_3CH(OH)CH_3$ Molar Mass: _____

                                                                     Polarity: _____

   Ethyl acetate $CH_3(CO)OCH_2CH_3$ Molar Mass: _____

                                                                     Polarity: _____

2. List the <u>complete set</u> of intermolecular forces for each liquid:

   Water   _____
   n-hexane   _____
   2-propanol   _____
   Ethyl acetate   _____

Materials Needed:
1. 10 mL graduated cylinder, 400 mL beaker, and five 10 mL test tubes
2. 50 mL side arm test tube with one-hole rubber stopper, thermometer, a support stand, and a clamp
3. Distilled water, and n-hexane, 2-propanol, ethyl acetate—about 50 mL of each
4. A mass balance and a calculator

EXPERIMENTAL PROCEDURE

1. **Determine the density of each liquid:**

   Use a 10 mL graduated cylinder. Weigh it and get approximately 5 mL of the first liquid. Record the volume accurately, and weigh. Since this is a liquid, you will need to use the weight by a difference technique to get the mass of the liquid. For each liquid, record your measured values and calculated density in the table below:

   Table 1. Density of various liquids at room temperature

	Water	n-Hexane	2-Propanol	Ethyl Acetate
Volume (mL)				
Mass of graduated cylinder (g)				
Mass of graduated cylinder + liquid (g)				
Mass of liquid (g)				
Density (g/mL)				

2. **Determine the boiling point of each liquid:**

   Pour the 5 mL sample from your graduated cylinder into a "side-arm" test tube. Insert a thermometer into a one-hole rubber stopper and close your side arm test tube with this thermometer-equipped rubber stopper. Fill a 400 mL beaker (heating bath) to its ¾ volume capacity with water and heat it up on a hot plate. Immerse the "side-arm" test tube containing the liquid so that the liquid surface is below liquid level in the heating bath. The tip of the thermometer should be above the liquid so that it can measure the temperature of the vapor when the liquid boils.

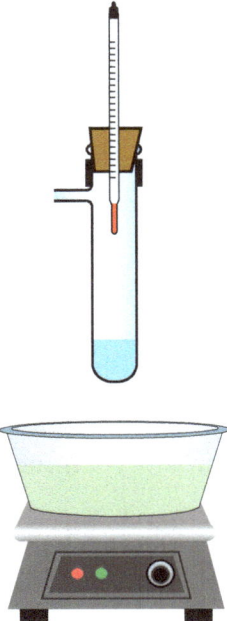

Watch the thermometer carefully as you heat the water. The temperature will rise until your sample starts to boil or the vapor will start to condense on the bulb of the thermometer. This is the boiling point of your liquid, and it will not rise further. You should recognize that boiling and condensing are opposite phase changes, but they happen at the same temperature. Once your sample is condensing on the bulb of the thermometer and dripping back into the solution, the temperature reading should stabilize at a steady state value. This means the thermometer is now reading at the boiling point of the liquid. Record the boiling point of your liquid. Remove the test tube and cool it down by running cold tap water on it. Repeat the measurement of boiling point two more times.

Table 2. Boiling point temperature of various liquids

	Boiling Point Measurements			Average Boiling Point	
	1	2	3	(°C)	(K)
n-Hexane					
2-Propanol					
Ethyl Acetate					

3. **Determine the miscibility of a liquid:**

   Select one of the liquids as your sample liquid: _____

   Test if your sample liquid mixes well with each of the other liquids, and with distilled water: Take 4 small test tubes, making sure they are clean and completely dry. Add 1 mL of your liquid to each test tube. Now add 1 mL of de-ionized water to one test tube, and 1 mL of one liquid to the next, and the other liquid to the other test tube until you have mixed each liquid with your sample liquid. Cork the tubes and shake them vigorously to fully mix the liquids. Allow the test tubes to sit undisturbed for four minutes and then observe and record the results as **miscible** (one liquid phase throughout), **immiscible** (a boundary is visible showing two distinct layers with equal volumes), or **partially miscible** (the mixture is cloudy, or you see two layers of unequal volumes).

Table 3. Miscibility of various liquids

	Chosen Liquid: _____ miscibility with …
Water	
n-Hexane	
2-Propanol	
Ethyl acetate	

4. **Analyze your experimental data:**

   a) Is density related to **intermolecular forces** (IMFs) at all? Comment on whether density (g/cm^3) applies at individual molecule level or at macroscopic, that is, bulk level.

   _____
   _____
   _____

   b) List your liquids according to the boiling points you found (low to high). Which liquid has relatively strongest **IMFs**? Comment on the type of IMFs this liquid has, and on the type of IMFs in the liquid with relatively weakest IMFs?

   _____
   _____
   _____

   c) Comment on the polarity of liquids that did not mix at all. Are they both polar, both nonpolar, or do they have different polarities? What makes molecules of the same kind stick together rather than stick to the other kind?

   _____
   _____
   _____

THESE ARE MY OWN NOTES
(from listening to lectures, watching YouTube videos, etc.)

Date: _____

Date: _____

Date: _____

Date: _____

I wish the professor could explain this: _____

Gases and Condensed Phases

Gases and Condensed Phases

Gases and Condensed Phases

WHAT IS THERMODYNAMICS?

Flow of Energy

Thermodynamics involves the study of change in internal energy of chemical species. Change in internal energy is equal to change in thermal energy of the system and amount of work done. For chemical reactions which do not involve gases or change in volume or pressure, the work done is negligible. In this course, we will not discuss the "work done" component.

Note that even the coldest item, relative to what is normal to us, has thermal energy. When we say "heat" in thermodynamics, we mean thermal energy. For this reason, we will not use degrees Celsius (°C) or Fahrenheit (°F) since these are

Looking at the pictures below, we know that dry ice (−79°C, left) is _____ than Ice (0°C). Is liquid water (25°C) hot? _____. If I put ice in dry ice it will _____ some of its internal energy. Its temperature will _____. If we wait long enough, ice outside the glass will _____ and its temperature will _____. But temperature of liquid water in the glass, to which ice is now added will _____. Is there a source of heat in this picture? _____ Explain. _____
_____

KPG_Payless/Shutterstock.com

© Africa Studio/Shutterstock.com

My Professor Says:

Do I get it? Let me check:

Initial | Later

Is his internal energy (right picture) the same as initial or did it change?
_____

subjective to human normalcy. We will use the Kelvin (K) scale. All chemical species which make matter have a positive Kelvin temperature and therefore internal energy. Temperature is really a measure of kinetic energy including vibrations of particles in any substance. It is practically impossible for any substance to reach 0 K. Furthermore, at 0 K there will not be any chemical activity. Chemical species will not interact—they will be dead. There will not be air to breathe and all life processes will cease. Since we cannot reach 0 K, we cannot determine absolute amount of internal energy of anything. We can only determine its change. Change in internal energy of chemical species is observed as heat gained or lost during a reaction where pressure and moles of gas (if any) are constant.

When this heat is measured at constant pressure it is called *enthalpy change of a reaction, ΔH_{rxn}*. We will also qualitatively discuss change in orderliness of a system (*entropy change, ΔS_{rxn}*), as well as spontaneity of chemical reactions based on *free energy change, ΔG_{rxn}*.

Change in enthalpy of a reaction (ΔH_{rxn}) is governed by the *first law of thermodynamics* which states that total energy in the universe is constant. The universe is simply the reacting chemical species and their surroundings. When the reacting species release energy (in the *exothermic* case), it is absorbed by the surroundings. Conversely, the reacting species absorb energy (in the *endothermic* case) from the surroundings.

My Professor Says:

Five ways of determining enthalpy change of a chemical reaction, ΔH_{rxn}, are:

1. Stoichiometry of a thermochemical equation [given mass of reactant or product predict/calculate the value of ΔH_{rxn} expected]
2. Calorimetry [$\mathbf{\Delta H_{rxn} = q = s\ m\ \Delta T}$, where $\mathbf{s}$ = specific heat capacity, $\mathbf{m}$ = mass, $\mathbf{\Delta T} = T_{final} - T_{initial}$]
3. Bond energies [$\Delta H°_{rxn} = \Sigma \Delta H°_{(reactant\ bonds\ broken)} + \Sigma \Delta H°_{(product\ bonds\ formed)}$]
4. Hess' law [manipulate known thermochemical equations to derive target equation]
5. Standard enthalpy of formation ($\Delta H_f°$)

Thermochemical Equations

Chemical equations which are written along with the amount of heat released or absorbed are called *thermochemical equations*. The energy term is stoichiometric. Let's explain this using an example:

$$N_2H_4(l) + 2H_2O_2(l) \rightarrow N_2(g) + 4H_2O(g)$$
$$\Delta H = -642\ kJ$$

This means 642 kJ is released during the reaction. We use the negative sign to indicate this. Exactly 642 kJ is produced from 1 mole of N_2H_4, or from 2 moles of H_2O_2, or along with 1 mole N_2, or 4 moles of H_2O. In short, we can relate the amount of heat produced to moles of any reactant or product. If you are asked "How much heat will be produced from 6.00 g H_2O_2?," you can answer by simply converting 6.00 g H_2O_2 to moles and multiplying by

Do I get it? Let me check:

Practice Questions

8.1. If 1.00 g of glucose produces 3.72 kcal, what is the ΔH value (in kJ) for the equation below? [1 cal = 4.184 J]
$C_6H_{12}O_6 + 6\ O_2 \rightarrow 6\ H_2O + 6CO_2$
I just need to convert 3.9 kcal/1.0 g to kJ/mol. Let me try:

My Professor Says:

642 kJ per 2 mol H_2O_2. Your calculation set-up should look like this:

$$6.00 \text{ g } H_2O_2 \times \frac{1 \text{ mol } H_2O_2}{34.02 \text{ g } H_2O_2} \times \frac{-642 \text{ kJ}}{2 \text{ mol } H_2O_2}$$
$$= -113 \text{ kJ}$$

In these calculations, the sign of the amount of heat simply indicates the direction of flow of heat (away from the system being negative and into the system being positive). What if we are asked to calculate the amount of H_2O_2 needed to produce 985 kJ? In the calculation set-up, we should attach the negative sign to 985 kJ since it is produced.

$$-985 \text{ kJ} \times \frac{2 \text{ mol } H_2O_2}{-642 \text{ kJ}} \times \frac{34.02 \text{ g } H_2O_2}{1 \text{ mol } H_2O_2}$$
$$= 104 \text{ g } H_2O_2$$

There are a few more things to note about thermochemical equations. When the physical state of reactants or products change, enthalpy change is affected. For example, when condensed water is produced instead of water vapor, 818 kJ is released:

$N_2H_4(l) + 2H_2O_2(l) \rightarrow N_2(g) + 4H_2O(l)$
$$\Delta H = -818 \text{ kJ}$$

When a thermochemical equation is reversed, enthalpy change takes on the opposite sign:

$N_2(g) + 4H_2O(l) \rightarrow N_2H_4(l) + 2H_2O_2(l)$
$$\Delta H = +818 \text{ kJ}$$

Do I get it? Let me check:

Practice Questions

8.2. Natural gas burns in air to form carbon dioxide and water, releasing heat, according to the following equation:

$CH_4(g) + 2O_2(g) \rightarrow CO_2(g) + 2H_2O(g)$
$$\Delta H_{rxn} = -802.3 \text{ kJ}$$

What is the maximum amount of heat that can be produced by 8.63 g of $CH_{4(g)}$? *After converting to mol, I can use 802.3 kJ/mol CH_4 to get the answer.*

My Professor Says:

An endothermic change requires heat as a reactant while exothermic change releases heat as product. Lastly, note that enthalpy change of a reaction is determined experimentally. So, if a thermochemical equation has fractions as coefficients such as those in molar decomposition of aluminum oxide, we do not convert them to whole numbers but use them as if they were whole numbers:

$Al_2O_3(s) \rightarrow 2Al(s) + \frac{3}{2}O_2(g)$
$\Delta H_{rxn} = 1676 \text{ kJ}$

Calorimetry

Calories (cal) are units of energy released by chemical species. We use a calorimeter to measure calories. For constant (atmospheric) pressure calorimetry, we put a magnetic stirrer in one polystyrene cup with a lid and nest it into another. We then insert a thermometer through the lid to make a calorimeter. We put reactant solutions in this calorimeter and allow the reaction to occur while we watch the temperature change on the thermometer.

After the maximum or minimum temperature has been reached, we use the calorimetry equation: $q = s\,m\,\Delta T$, where s is the specific heat capacity of the solvent (water, in most cases). *Specific heat capacity* is the amount of heat required to change 1 g of the substance by 1 K. Specific heat capacity of water is 4.184 J/g.K. Since 1 K increment is the same size as 1°C increment, units of J/g°C can also be used.

Do I get it? Let me check:

Practice Question

8.3. Given the equation: $q = s\,m\,(T_f - T_i)$, solve for T_f. *This is like solving for x. I got this:*

My Professor Says:

The other kind of calorimeter, where pressure may increase especially when gases are produced, is called constant volume calorimetry. Since the calorimeter itself absorbs heat released, the equation is simply **q = s ΔT**, where specific heat capacity of the calorimeter has units of J/K. If we can measure change in temperature of the substances releasing or gaining heat, we can do calorimetry.

Example 1: "Find ΔH_{rxn}, enthalpy change, for the reaction between Mg and HCl in 100.0 g aqueous solution (s = 4.18 J/g.K). The reaction raises temperature of the solution from 25.6°C to 32.8°C." Remember that the difference between °C-values and K-values is numerically the same since scale increments are of the same size. Also, notice that ΔH_{rxn} *is simply heat, q.*

$$\Delta H_{rxn} = q_{solution} = sm\Delta T$$

$$\Delta H_{rxn} = 4.184 \, J/g.K \times 100.0 \, g \times (32.8 - 25.6) K$$

$$= 3.01 \times 10^3 \, J$$

Example 2: "Find the specific heat capacity of a 22.05 g solid that was heated in a test-tube to 100.0°C. When it was added to 50.0 g of water (s = 4.18 J/g.K) in a coffee-cup calorimeter, water temperature changes from 25.10°C to 28.49°C."

Since water and the solid are in contact, heat is transferred from the solid to the water until they reach the same T_{final}. In addition, the heat given out by the solid ($-q_{solid}$) is equal to the heat absorbed by the water (q_{water}).

Do I get it? Let me check:

Practice Question

8.4. If 14.5 J of heat is absorbed by a 5.0 g Cu-plate (s = 0.385 J/g°C), from a hot weld, raising Cu temperature to 30.5°C, find the initial temperature of Cu-plate. [q = smΔT]. *This is like solving for x. I got this:*

My Professor Says:

$-q_{solid} = q_{water}; -(sm\Delta T)_{solid} = (sm\Delta T)_{water}$

$s_{solid} = \dfrac{(sm\Delta T)_{water}}{(m\Delta T)_{solid}}$

$s_{solid} = \dfrac{4.184 \, J/g \cdot K \times 50.0 \, g \times (28.49 - 25.10) \, K}{-[22.05 \, g \times (28.49 - 100.0) \, K]_{solid}}$

$= 0.450 \, J/g \cdot K$

Do I get it? Let me check:

Practice Question

8.5. Magnesium metal is added to $HCl_{(aq)}$ in a coffee-cup calorimeter to make 100.0 g solution:

$Mg(s) + 2HCl(aq) \rightarrow MgCl_2(aq) + H_2(g)$

Solution temperature rises from 25.6°C to 32.8°C. Find ΔH_{rxn}. [$s = 4.184 \, J/(g \cdot K)$].
I can use $q = sm\Delta T$:

Bond Energy

Another way of finding enthalpy of a reaction is using bond energies. Different bond energies have been determined experimentally and averages are listed in textbooks for reference. There are some basics that we must remember:

1. Since it takes energy to do, breaking bonds is endothermic. In a reaction, bonds of reactants are always broken.

2. Energy of combined species is always less than the sum of energies of individual species. So, forming bonds is exothermic. In a reaction, bonds of products are always formed.

3. The side that has higher energy determines whether a chemical reaction will be endothermic or exothermic.

The equation that we use to calculate enthalpy change of a reaction (ΔH_{rxn}) based on bond energies is shown below:

$\Delta H°_{rxn} = \Sigma \Delta H°_{(bonds\ broken)} + \Sigma \Delta H°_{(bonds\ formed)}$

The energies of individual bonds are shown in the table. Draw Lewis structures of each reactant and each product in

My Professor Says:

the chemical equation—so you can see the single, double, and triple bonds. Then add up the bond energies of each reactant multiplied by its balancing coefficient. Do the same for products. Then plug the sum of reactant bonds broken as a positive value, and the sum of product bonds formed as a negative value.

Example: Use bond energies in the table to calculate $\Delta H°_{rxn}$ for the chlorination of methane to form chloroform:

Bond:	C–H	Cl–Cl	C–Cl	H–Cl
Energy (kJ/mol):	413	243	339	427

$$H_3C\text{–}H + 3\, Cl\text{–}Cl \longrightarrow CHCl_3 + 3\, H\text{–}Cl$$

Notice that all the reactant bonds break, and all the product bonds form. Calculate and substitute the two sums, with correct signs, into the $\Delta H°_{rxn}$ equation.

Bonds broken	Bonds formed
4 × C-H	3 × C-Cl
3 × Cl-Cl	1 × C-H
	3 × H-Cl
$\Sigma \Delta H°$: = 2381 kJ	= −2711 kJ

$\Delta H°_{rxn} = \Sigma \Delta H°_{\text{(bonds broken)}} + \Sigma \Delta H°_{\text{(bonds formed)}}$

$= 2381 \text{ kJ} + (-2711 \text{ kJ}) = -330\ kJ$

Do I get it? Let me check:

Practice Question

8.6. Use bond energies below to calculate the enthalpy change (ΔH, in kJ) for the following reaction:

$$:C\equiv O: + H_3C\text{–}O\text{–}H \longrightarrow H_3C\text{–}C(=O)\text{–}O\text{–}H$$

Bond:	C–C	C≡O	C–H	O–H	C–O	C=O
B.E. (kJ/mol)	347	1070	413	467	358	745

My Professor Says:

Hess' Law

A scientist named Germain Hess figured out that, since thermochemical equations of many chemical reactions are listed in reference books, we can simply manipulate them along with the enthalpy term (add, reverse, or multiply them) to derive the equation we want and its enthalpy change.

Here is a chemical equation of interest to an environmental chemist because it converts noxious gases into green gases, that is, those readily used by animals and plants:

CO(g) + NO(g) → CO$_2$(g) + ½N$_2$(g)
$$\Delta H_{rxn}=?$$

Some thermochemical equations in reference books which we can manipulate to get the one our environmental chemist is studying are:

1. CO(g) + ½ O$_2$(g) → CO$_2$(g)
$$\Delta H_1 = -283.0 \text{ kJ}$$

2. N$_2$(g) + O$_2$(g) → 2NO(g)
$$\Delta H_2 = 180.6 \text{ kJ}$$

According to Hess, we simply check if reactants and products of the given equations align with the target equation. Equation 1 seems to be aligned since CO and CO$_2$ are on the correct sides, respectively. Equation 2 is not aligned. We must flip it around so that NO is on the reactant side and N$_2$ is on the product side. *When we flip or reverse the equation, the sign of ΔH changes*. We must also *multiply* each term in *Equation 2 by ½, including ΔH*.

Do I get it? Let me check:

Practice Question

8.7. Given the following chemical equations:

a) C(s) + O$_2$(g) → CO$_2$(g)
$$\Delta H°_{rxn} = -393.5 \text{ kJ}$$

b) 2CO(g) + O$_2$(g) → 2CO$_2$(g)
$$\Delta H°_{rxn} = -566.0 \text{ kJ}$$

c) 2H$_2$(g) + O$_2$(g) → 2H$_2$O (g)
$$\Delta H°_{rxn} = -483.6 \text{ kJ}$$

Calculate $\Delta H°_{rxn}$ for the reaction below:

C(s) + H$_2$O(g) → CO(g) + H$_2$(g)
$$\Delta H°_{rxn} = \underline{}$$

My Professor Says:

Reverse Equation 2 and multiply by ½:

NO(g) → ½ N$_2$(g) + ½ O$_2$(g); $\Delta H = -90.3$ kJ

Now that we have aligned the given equations in terms of positions and coefficients, we can add the two and see if we get the target equation. To find ΔH of the target equation, we add the ΔH values we obtained after manipulating each given equation as follows:

Equation 1 + Equation 2 (modified):

CO(g) + NO(g) → CO$_2$(g) + ½ N$_2$(g)

$\Delta H_{rxn} = \Delta H_1 + \Delta H_2 \text{(modified)} = -373.3$ kJ

Standard Enthalpies of Formation (ΔH_f°)

The thinking on this method of finding ΔH_{rxn} is that every single compound is in its standard state and can be made from the individual elements. Standard state is the form in which the compound is found at 298 K and 1 atm. We use fractional coefficients if necessary because we must form exactly 1 mol of a given compound. The elements may be in their most stable allotropic or natural form which will have enthalpy of formation of "0." Other allotropes or forms will have relatively higher enthalpies of formation. Nevertheless, our job is to write chemical equations of elements forming 1 mol of a compound. For every species in the target equation, if the ΔH_f of the compound is not already listed. We can then use these ΔH_f values to calculate ΔH_{rxn}.

Do I get it? Let me check:

Practice Question

8.8. Write a balanced equation for the formation of 1 mol of Al$_2$(SO$_4$)$_3$ ($\Delta H^\circ_f = -3440$ kJ/mol)

My Professor Says:

Example: Write a balanced equation for the formation of 1 mol of HCN(g) with $\Delta H°_f = 135$ kJ.

$$C(graphite) + \tfrac{1}{2}N_2(g) + \tfrac{1}{2}H_2(g) \rightarrow HCN(g)$$
$$\Delta H_f° = 135 \text{ kJ}$$

In many cases ΔH_f of many compounds are listed in reference tables. The equation we use to find change in enthalpy of a chemical reaction is:

$\Delta H°_{rxn} = \Sigma \Delta H°_{(products)} + \Sigma \Delta H°_{(reactants)}$

Example: Use $\Delta H_f°$ values in the table to calculate ΔH_{rxn} for the oxidation of ammonia:

	NO(g)	H$_2$O(g)	NH$_3$(g)	O$_2$(g)
ΔH_f (kJ/mol)	90.3	−241.8	−45.9	0

$4NH_3(g) + 5O_2(g) \rightarrow 4NO(g) + 6H_2O(g)$

Products	Reactants
4 × NO(g)	4 × NH$_3$(g)
6 × H$_2$O(g)	5 × O$_2$(g)

$\Sigma\Delta H°: = -1089.6$ kJ = 183.6 kJ

$\Delta H°_{rxn} = \Sigma\Delta H°_{(products)} - \Sigma\Delta H°_{(reactants)}$
   $= -1089.6 \text{ kJ} - 183.6 \text{ kJ}) = $ *−906 kJ*

Entropy Change (ΔS_{rxn})

The *second law of thermodynamics* says a spontaneous change goes in the direction that increases entropy. By spontaneous we mean that the process occurs without continuous supply of energy from the surroundings. One example of spontaneous processes is a melting block of ice.

Do I get it? Let me check:

Practice Question

8.9. Use $\Delta H_f°$ values in the table to calculate ΔH_{rxn} for the oxidation of glucose:

	C$_6$H$_{12}$O$_6$(s)	O$_2$(g)	H$_2$O(l)	CO$_2$(g)
ΔH_f (kJ/mol)	−1273.3	0	−285.8	−393.5

$C_6H_{12}O_6(s) + 6\,O_2(g) \rightarrow 6\,H_2O(l) + 6CO_2(g)$

I can use $\Delta H°_{rxn} = \Sigma\Delta H°_{(products)} + \Sigma\Delta H°_{(reactants)}$

Entropy change is nature's tendency to disperse energy. To better understand change in entropy, we imagine that every system has microstates like energy levels. Various factors may contribute to entropy of a system:

1. Temperature which indicates motion of particles of a substance

2. The physical state of a substance. Gas has more energetic particles than liquid and solid)

3. A dissolved compound has higher entropy than when it is undissolved

4. Particle size and complexity also indicate the level of ΔS

It is very important to develop a sense of change in entropy in order to predict a positive or negative entropy change.

Practice Question

8.10. For each process indicate whether the change in entropy (ΔS) is **positive** or **negative**.

a) $4NH_3(g) + 5O_2(g) \rightarrow 4NO(g) + 6H_2O(g)$
   $\Delta S°_{rxn}$ is _____

b) $2\,O_3(g) \rightarrow 3\,O_2(g)$
   $\Delta S°_{rxn}$ is _____

c) $C_3H_8(g) + 5O_2(g) \rightarrow 3CO_2(g) + 4H_2O(l)$
   $\Delta S°_{rxn}$ is _____

d) $H_2C\underset{\diagdown}{\overset{CH_2}{-}}CH_2\,(g) \xrightarrow{\Delta} CH_3-CH=CH_2\,(g)$
   $\Delta S°_{rxn}$ is _____

My Professor Says:

Free Energy Change, ΔG_{rxn}

The *free energy change* (ΔG) is a measure of energy which allows change to occur spontaneously, and it is given by Gibbs' equation:

$$\Delta G = \Delta H - T\Delta S.$$

When we know the sign of enthalpy change (ΔH) and that of entropy change (ΔS), we can plug them into the Gibbs' equation to see what sign ΔG will have. A negative ΔG indicates that a reaction will occur spontaneously. A positive ΔG indicates nonspontaneity. A zero ΔG implies that the system is at equilibrium and there is no net change.

Do I get it? Let me check:

Practice Question

8.11. The decomposition of water molecules into gaseous oxygen and hydrogen molecules by electrolysis is shown below.

a) Is the sign of ΔH_{sys} for this process negative or positive? _____.
   Explain: _____
   _____

b) Is the sign of ΔS_{sys} for this process negative or positive? _____.
   Explain: _____
   _____

c) This process is spontaneous when sign of ΔG_{sys} is _____, which will occur at what temperature level?
   _____

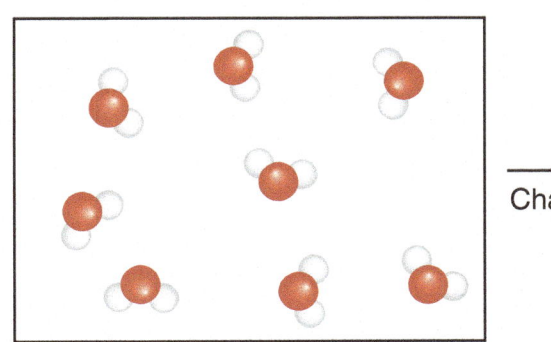
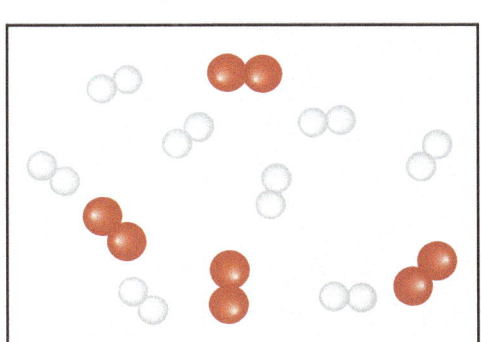

Electrolysis of water

My Professor Says:

Do I get it? Let me check:

Knowing that a given chemical reaction occurs spontaneously if $\Delta G < 0$, but it is nonspontaneous if $\Delta G > 0$ we can deduce from Gibbs' equation, $\Delta G = \Delta H - T\Delta S$, that:

- If $\Delta H < 0$ and $\Delta S > 0$, then $\Delta G < 0$ for all T. So, the reaction is spontaneous.

- If $\Delta H > 0$ and $\Delta S < 0$, then $\Delta G > 0$ for all T. So, the reaction is nonspontaneous.

- If $\Delta H > 0$ and $\Delta S > 0$, then ΔG becomes more negative, that is, the reaction becomes more spontaneous, as T increases.

- If $\Delta H < 0$ and $\Delta S < 0$, then ΔG becomes more negative, that is, the reaction becomes more spontaneous, as T decreases.

ΔH	ΔS	ΔG	Spontaneous?
−	+	$\Delta G = (-) - [T\,(+)] = -$	**Always**, regardless of T
+	−	$\Delta G = (+) - [T\,(-)] = +$	**Never**, regardless of T
+	+	$\Delta G = (+) - [T\,(+)] = ?$	spontaneous at **high** T, where $\Delta G < 0$
−	−	$\Delta G = (-) - [T\,(-)] = ?$	Spontaneous at **low** T, where $\Delta G < 0$

THESE ARE MY OWN NOTES
(from listening to lectures, watching YouTube videos, etc.)

Date: _____

Date: _____

Date: _____

Date: _____

I wish the professor could explain this: _____

Student's Practical Laboratory Experiment

MEASURING CHANGE IN ENERGY

OBJECTIVE

The objective of this experiment is to determine change in enthalpy (ΔH) of both physical and chemical reactions.

During dissolution of an ionic compound the ions, which are bonded to each other electrostatically forming ionic bonds, are being pulled apart, and new attractions between water molecules and the ions, called intermolecular forces (not bonding forces), occur. The overall sum of the *energy required to break ionic bonds (always a "+" value)* and *energy released when new attractions form (always a "−" value)* will determine if additional energy must be put in to make the dissolution happen, or if energy will be released during the process.

In a chemical reaction, the breaking of chemical bonds requires input of energy, and forming new chemical bonds gives off energy. The sum of the energies of the bonds that are broken in a reaction and the new bonds that are formed determine the overall change in energy.

In this lab, we will measure the heat change in these two processes using a simple calorimeter. We will then calculate how much heat was either generated or absorbed and make it part of a balanced equation.

LEARNING OUTCOMES

After this experiment, students will be able to:

1. Evaluate enthalpy change (ΔH) for dissolution of ionic compound, and a chemical reaction
2. Draw Energy versus Progress diagram for a physical and chemical change
3. Write a thermochemical equation to represent specific chemical interactions

SAFETY NOTES

The reaction between Mg and HCl produces fumes. Work in a fume hood or next to the air vent to avoid inhaling fumes. Hydrochloric acid solution is corrosive. Upon any skin contact, rinse with plenty of water. Always wear safety goggles and shoes when doing experiments. *When finished, wash your hands with soap and water.*

PRE-LAB ASSIGNMENT

In *Part I*, you will study the dissociation of baking soda. The chemical name for baking powder (or baking soda) is sodium bicarbonate, also called sodium hydrogen carbonate. It is an ionic solid, and its chemical formula is $NaHCO_3$. Write the equation for its dissolution. Be sure to include charges and physical states.

$$NaHCO_3(s) \rightarrow \underline{\hspace{2cm}} + \underline{\hspace{2cm}}$$

In *Part II*, you will study the reaction between magnesium metal and hydrochloric acid $HCl(aq)$. The products of this reaction are hydrogen gas and the ionic compound magnesium chloride, which remains dissolved in solution. From this description, write a balanced molecular equation for this reaction. Make sure you include physical states for all the chemical species.

$$\underline{\hspace{2cm}} + \underline{\hspace{2cm}} \rightarrow \underline{\hspace{2cm}} + \underline{\hspace{2cm}}$$

MATERIALS NEEDED

1. A mass balance weighing boat or paper, distilled water
2. A 250 mL beaker, a 25 mL pipet, and a 50 mL graduated cylinder
3. A styrofoam cup with lid, and a digital thermometer. This is your calorimeter. See picture.
4. A stopwatch [on your cellphone]
5. Baking powder, $NaHCO_3(s)$
6. Hydrochloric acid, 1.0 *M* HCl, solution
7. Solid magnesium, Mg

© focal point/Shutterstock.com

© Viktor Chursin/Shutterstock.com

EXPERIMENTAL SECTION

Part I. Baking Powder Dissolution

1. Tare balance and take mass of your calorimeter, without lid and thermometer.
2. Use your 25 mL pipet to measure out 50.0 mL of DI water into the calorimeter cup.
3. Tare balance and take total mass of "water + calorimeter."
4. Insert thermometer through the lid of your calorimeter and record water temperature.
5. Place calorimeter into a 250 mL beaker so it does not tip over.
6. Tare (zero) boat weight on the balance. Weigh out 1($\pm$ 0.005) g of baking powder. Record.
7. Start stopwatch. Gently transfer baking powder into calorimeter and take temperature at 20 second mark. Swirl the solution gently to ensure good mixing.
8. Take temperature readings (in °C) every 20 seconds over a total time of 240 seconds.
9. Tare balance and take mass of "Final Solution + Calorimeter."
10. Carefully pour the solution into a 50.0 mL graduated cylinder. Get as much of the solution into the cylinder as you can. Record the volume of the entire solution.

	Readings (w units)
1. Mass of the "Calorimeter"	
2. Mass of the "Calorimeter + Mass of 50.00 mL water"	
3. Temperature of the water (before $NaHCO_3$) – "0" second	
4. Mass of $NaHCO_3$ [Start the stopwatch]	
5. Add $NaHCO_3(s)$ and take Temperature at "0 + 20" second	
–Temperature at 40 seconds	
–Temperature at 60 seconds	
–Temperature at 80 seconds	
–Temperature at 100 seconds	
–Temperature at 120 seconds	
–Temperature at 140 seconds	
–Temperature at 160 seconds	
–Temperature at 180 seconds	
–Temperature at 200 seconds	
–Temperature at 220 seconds	
–Temperature at 220 seconds	
6. Mass of "Final Solution + Calorimeter"	
7. Volume of "final solution," $NaHCO_3(aq)$	

Part II. "Mg(s) + HCl(aq)" Reaction

1. Rinse the styrofoam cup with distilled water. Wipe it dry with a paper towel.
2. Rinse the pipet with distilled water and then with 1.0 M HCl solution.
3. Use the pipet to measure out 25 mL of HCl solution into the styrofoam cup.
4. Weigh (and record mass of) the styrofoam cup +1.0 M HCl solution.
5. Insert thermometer through the lid of your calorimeter. Record HCl solution temperature.
6. Obtain a 0.5 g piece of magnesium from your instructor. Tare balance and take mass.
7. Start stopwatch.

8. Slide Mg metal gently into calorimeter and take temperature at 20s second mark. Make sure Mg metal is immersed. Swirl the solution gently to ensure good mixing.
   Caution: Work next to air vent or in the fume hood to avoid inhaling fumes.
9. Take v readings (in °C) every 20 seconds over a total time of 240 seconds.
10. Tare balance and take mass of "Final Solution + Calorimeter."
11. Carefully pour the solution into a 50.0 mL graduated cylinder. Get as much of the solution into the cylinder as you can. Record the volume of the entire solution.

	Readings (w units)
1. Mass of the "Calorimeter" from Part I.	
2. Mass of "Calorimeter + Mass of 25.0 mL HCl solution"	
3. Temperature of 1.0 M HCl solution (before Mg) – "0" sec	
4. Mass of Mg metal [Start the stopwatch]	
5. Add Mg metal and take Temperature at "0 + 20" seconds	
–Temperature at 40 seconds	
–Temperature at 60 seconds	
–Temperature at 80 seconds	
–Temperature at 100 seconds	
–Temperature at 120 seconds	
–Temperature at 140 seconds	
–Temperature at 160 seconds	
–Temperature at 180 seconds	
–Temperature at 200 seconds	
–Temperature at 220 seconds	
–Temperature at 220 seconds	
6. Mass of "Final Solution + Calorimeter"	
7. Volume of "final solution," $MgCl_2(aq)$	

DATA ANALYSIS

1. Fill in the table (include the sign) to summarize your results. To evaluate enthalpy changes (ΔH), use specific heat capacity of water since the solvent is the overwhelming component of the final mixture [**s** of H_2O = 4.184 J/g.°C].

	Part I	Part II
Mass of Final solution, *m*		
Final Temperature, T_2	°C	°C
Initial Temperature, T_1	°C	°C
Temperature Change, ΔT ($\Delta T = T_2 - T_1$)	°C	°C
Enthalpy Change (ΔH) ($\Delta H = q = sm\Delta T$)		

2. Draw **Energy vs. Progress diagrams** for "$NaHCO_3(s)$ dissolution" and "$Mg(s) + HCl(aq)$ reaction," respectively.

   a) <u>Part I.</u> Baking Powder Dissolution
   Notice that the Temperature Change you observed, from *initial* to *final* states, is that of solvent molecules which surround $NaHCO_3$ granules. If temperature dropped, the energy was absorbed by the reactants, from solvent molecules. If you touched the solution, energy was drawn from your hand and got colder. Draw the **Energy *vs.* Progress** diagram to show change in energy during solution formation.

b) <u>Part II.</u> "Mg(s) + HCl(aq)" Reaction
Again, notice that the Temperature Change you observed, from *initial* to *final* states, is that of solvent molecules which surround Mg and H$^+$ ions (from HCl) which exchange electrons. As we discussed before, Mg donates two electrons (2e$^-$), one to each H$^+$ ions which then form H$_2$ molecules and releasing energy to the surrounding solvent molecules. If temperature increased, energy was absorbed by solvent molecules from the reactants. Draw the **Energy *versus* Progress** diagram to show change in energy during the reaction.

3. Now that we know the amount of energy transferred in each reaction, we can put it into the balanced equation. To do that, we need to put it on a per mole basis, or J/mol of one of the reactants. We will use the solid reactants to do this in the following steps, because these were the limiting reactants in each reaction. (We know this without calculating because they were completely consumed!)

	Part I	Part II
a) Convert mass of each solid reactant into moles		
b) Divide the total energy in kJ for each reaction by the moles of reactant consumed.		
c) The answer is the amount of energy produced per mole of reactant. Multiply this number by the coefficient in the balanced equation for that reactant.		

4. Now write the balanced equation with the balanced energy term on the correct side of the equation.

   a) <u>Part 1:</u>

   b) <u>Part 2:</u>

—o0o—   **END OF LAB REPORT**

THESE ARE MY OWN NOTES
(from listening to lectures, watching YouTube videos, etc.)

Date: _____

Date: _____

Date: _____

Date: _____

I wish the professor could explain this: _____

What is Thermodynamics?